Defect

Detect

Linux API
for Software Diagnostics
Accelerated

With Category Theory in View

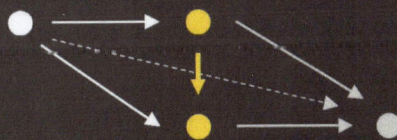

Dmitry Vostokov
Software Diagnostics Services

Published by OpenTask, Republic of Ireland

OpenTask books and magazines are available through booksellers and distributors worldwide. For further information or comments, send requests to press@opentask.com.

A CIP catalog record for this book is available from the British Library.

ISBN-l3: 978-1-912636-62-4 (Paperback)

Revision 1.02 (July 2023)

Contents

About the Author

Dmitry Vostokov is an internationally recognized expert, speaker, educator, scientist, inventor, and author. He is the founder of the pattern-oriented software diagnostics, forensics, and prognostics discipline (Systematic Software Diagnostics), and Software Diagnostics Institute (DA+TA: DumpAnalysis.org + TraceAnalysis.org). Vostokov has also authored more than 50 books on software diagnostics, anomaly detection and analysis, software and memory forensics, root cause analysis and problem solving, memory dump analysis, debugging, software trace and log analysis, reverse engineering, and malware analysis. He has over 25 years of experience in software architecture, design, development, and maintenance in various industries, including leadership, technical, and people management roles. Dmitry also founded Syndromatix, Anolog.io, BriteTrace, DiaThings, Logtellect, OpenTask Iterative and Incremental Publishing (OpenTask.com), Software Diagnostics Technology and Services (former Memory Dump Analysis Services) PatternDiagnostics.com, and Software Prognostics. In his spare time, he presents various topics on Debugging.TV and explores Software Narratology, its further development as Narratology of Things and Diagnostics of Things (DoT), Software Pathology, and Quantum Software Diagnostics. His current interest areas are theoretical software diagnostics and its mathematical and computer science foundations, application of formal logic, artificial intelligence, machine learning and data mining to diagnostics and anomaly detection, software diagnostics engineering and diagnostics-driven development, diagnostics workflow and interaction. Recent interest areas also include cloud native computing, security, automation, functional programming, applications of category theory to software diagnostics, development and big data, and diagnostics of artificial intelligence.

Introduction

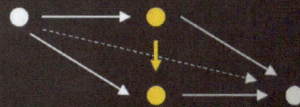

Linux API
for Software Diagnostics
Accelerated
With Category Theory in View

Dmitry Vostokov
Software Diagnostics Services

Hello everyone, my name is Dmitry Vostokov, and I teach this training course.

Prerequisites

- Development experience

and (optional)

- Basic process core dump analysis

To get most of this training, you are expected to have basic development experience and optional basic process core dump analysis experience. I assume you know what types, functions, and their parameters are. If you don't have a core dump analysis experience, then you also learn some basics too because we use GDB and optionally the Microsoft debugger, WinDbg (classic) from Debugging Tools for Windows, or the WinDbg app (former WinDbg Preview) for some exercises. I explain some debugging and related concepts when necessary during the course.

Training Goals

- Review fundamentals of Linux API

- Learn diagnostic analysis techniques

- See how Linux API knowledge is used during diagnostics and debugging

Our primary goal is to learn Linux API in an accelerated fashion. So, first, we review Linux API fundamentals necessary for software diagnostics. Then we learn various analysis techniques for Linux API exploration. And finally, we see examples of how the knowledge of Linux API helps in diagnostics and debugging.

Training Principles

- ◉ Talk only about what I can show

- ◉ Lots of pictures

- ◉ Lots of examples

- ◉ Original content and examples

There were many training formats to consider, and I decided that the best way is to concentrate on slides and hands-on demonstrations you can repeat yourself as homework. Some of them are available as step-by-step exercises.

Schedule

- Review of relevant x64 and A64 disassembly

- General Linux API aspects

- Linux API formalization

- Linux API and languages

- Linux API classes

- Practical exercises

The rough coverage or schedule includes general API aspects that can also be applicable to other operating systems. We also take a radical detour and introduce category theory in the API context. Our coverage is not only theoretical. We also do a tour through different API subsets and classes. An integral part of this training is practical exercises.

Training Idea

- ◉ **Previous Accelerated Windows API training**

  - Cybersecurity

  - Memory dump analysis

  - Reading Windows-based Code training

- ◉ **Experience writing Linux API monitoring tools**

This training idea came from the previous Windows API training for security professionals who mentioned the need for Windows API knowledge and attendees of my memory dump analysis training courses who asked questions related to Windows API. I realized that since I have the Linux core dump analysis course, attendees of it would also benefit from similar training for Linux API too. This training may also fill some gaps from other training courses, such as Linux disassembly and reversing. Additional push came from my experience designing and implementing Linux API monitoring tools from the ground up.

General Linux API Aspects

- Header view
- Naming convention
- Basic type system
- Call types
- Export/import functions
- PLT and GOT
- Virtual process address space
- Calling convention
- API sequences
- API layers
- API and system calls
- API source code
- Shared Libraries
- API usage
- API internals

- Static linking
- Delayed dynamic linking
- API name patterns
- API namespaces
- API syntagms/paradigms
- Marked API
- ADDR patterns
- DebugWare patterns
- Memory analysis patterns
- API tracing
- Trace and log analysis patterns
- API and errors
- API and functional programming
- API and security
- API and versioning

The general Linux API aspects we plan to discuss are listed on this slide.

Linux API Formalization

- API compositionality
- Category theory language
- A view of category theory
- Category theory square
- API category
- API functor
- API diagram
- API natural transformation
- Cross-platform API
- API adjunction
- Informal n-API
- API and trace categories
- API I/O

This training also includes an API formalization via Category Theory. We start with a brief overview of categories, functors, and other aspects using Linux API examples.

Linux API and Languages

- C#
- Scala Native
- Golang
- Rust
- Python

We also cover the basics of how Linux API is used in languages other than C/C++ with template examples.

Linux API Classes

- System configuration
- File I/O
- File control
- Filesystem
- Dynamic memory
- Virtual memory
- Shared libraries
- Process
- IPC

- Job
- Signals
- Thread
- Networking
- Time
- Timers
- Tracing and logging
- Accounts
- Terminal

A part of this training includes a tour of different Linux API subsets and classes.

Links

- ⊙ Core Dumps

Included in Exercise L0

- ⊙ Exercise Transcripts

Included in this book

Exercise L0

- **Goal:** Install GDB and check if GDB loads a core dump correctly

- **Goal:** Install WinDbg or Debugging Tools for Windows, or pull Docker image, and check that symbols are set up correctly

- **Memory Analysis Patterns:** Stack Trace; Incorrect Stack Trace

- \LAPI-Dumps\Exercise-L0-GDB.pdf

- \LAPI-Dumps\Exercise-L0-WinDbg.pdf

For practical exercises, we need mostly needed GDB and its multiarchitecture version. If you are coming from a Windows background, you may try the same exercises with either the WinDbg app (former Windows Preview), the classical Microsoft WinDbg debugger from Debugging Tools for Windows, or even the Docker version. I decided to add WinDbg because of my recent experience with one critical software incident where GDB was not showing correct information but loading the same core dump in WinDbg helped. Here I assume you already prepared the environment and only show the multiarch GDB part, which I added for the first time to my Linux training courses.

Exercise L0 (GDB)

Goal: Install GDB and check if GDB loads a core dump correctly.

Memory Analysis Patterns: Stack Trace; Incorrect Stack Trace.

1.   Download core dump files if you haven't done that already and unpack the archives:

https://www.patterndiagnostics.com/Training/LAPI/LAPI-Dumps.zip

2.   Download and install the latest version of GDB. For WSL2 Debian, we used the following commands:

```
$ sudo apt install build-essential
$ sudo apt install gdb
```

To analyze A64 dumps on the same x64 host platform, we also installed multiarch GDB via:

```
$ sudo apt install gdb-multiarch
```

On our RHEL-type system, we installed the tools and GDB via:

```
$ sudo yum group install "Development Tools"
$ sudo yum install gdb
```

3.   Verify that GDB is accessible and then exit it (**q** command):

```
$ gdb
GNU gdb (Debian 8.2.1-2+b3) 8.2.1
Copyright (C) 2018 Free Software Foundation, Inc.
License GPLv3+: GNU GPL version 3 or later <http://gnu.org/licenses/gpl.html>
This is free software: you are free to change and redistribute it.
There is NO WARRANTY, to the extent permitted by law.
Type "show copying" and "show warranty" for details.
This GDB was configured as "x86_64-linux-gnu".
Type "show configuration" for configuration details.
For bug reporting instructions, please see:
<http://www.gnu.org/software/gdb/bugs/>.
Find the GDB manual and other documentation resources online at:
    <http://www.gnu.org/software/gdb/documentation/>.

For help, type "help".
Type "apropos word" to search for commands related to "word".

(gdb) q
$
```

4.   Load *core.9* dump file and *bash* executable from the x64/ directory:

```
$ cd LAPI/x64

~/LAPI/x64$ gdb -c core.9 -se bash
```

```
GNU gdb (Debian 8.2.1-2+b3) 8.2.1
Copyright (C) 2018 Free Software Foundation, Inc.
License GPLv3+: GNU GPL version 3 or later <http://gnu.org/licenses/gpl.html>
This is free software: you are free to change and redistribute it.
There is NO WARRANTY, to the extent permitted by law.
Type "show copying" and "show warranty" for details.
This GDB was configured as "x86_64-linux-gnu".
Type "show configuration" for configuration details.
For bug reporting instructions, please see:
<http://www.gnu.org/software/gdb/bugs/>.
Find the GDB manual and other documentation resources online at:
     <http://www.gnu.org/software/gdb/documentation/>.

For help, type "help".
Type "apropos word" to search for commands related to "word"...
Reading symbols from bash...(no debugging symbols found)...done.

warning: core file may not match specified executable file.
[New LWP 9]
Core was generated by `-bash'.
#0  0x00007f3e9f7492d7 in __GI__waitpid (pid=-1, stat_loc=0x7ffcbd661ad0, options=10) at
../sysdeps/unix/sysv/linux/waitpid.c:30
30          ../sysdeps/unix/sysv/linux/waitpid.c: No such file or directory
```

5.   Verify that the stack trace (backtrace) is shown correctly with symbols:

```
(gdb) bt
#0  0x00007f3e9f7492d7 in __GI__waitpid (pid=-1, stat_loc=0x7ffcbd661ad0, options=10) at
../sysdeps/unix/sysv/linux/waitpid.c:30
#1  0x000055e1650a4869 in ?? ()
#2  0x000055e1650a5cc3 in wait_for ()
#3  0x000055e165093b85 in execute_command_internal ()
#4  0x000055e165093df2 in execute_command ()
#5  0x000055e16507b833 in reader_loop ()
#6  0x000055e16507a104 in main ()
```

Note: If the stack trace on your system is incorrect, check that shared libraries are loaded, and if not, add the current directory to the shared library search path (see the A64 example below).

6.   We exit GDB.

```
(gdb) q
```

~/LAPI/x64$

7.   Load *core.19649* dump file and *bash* executable from the A64/ directory using multiarch GDB:

```
$ cd ../A64

~/LAPI/A64$ gdb-multiarch -c core.19649 -se bash
```

```
GNU gdb (Debian 8.2.1-2+b3) 8.2.1
Copyright (C) 2018 Free Software Foundation, Inc.
License GPLv3+: GNU GPL version 3 or later <http://gnu.org/licenses/gpl.html>
This is free software: you are free to change and redistribute it.
There is NO WARRANTY, to the extent permitted by law.
Type "show copying" and "show warranty" for details.
This GDB was configured as "x86_64-linux-gnu".
```

```
Type "show configuration" for configuration details.
For bug reporting instructions, please see:
<http://www.gnu.org/software/gdb/bugs/>.
Find the GDB manual and other documentation resources online at:
    <http://www.gnu.org/software/gdb/documentation/>.

For help, type "help".
Type "apropos word" to search for commands related to "word"...
Reading symbols from bash...(no debugging symbols found)...done.

warning: core file may not match specified executable file.
[New LWP 19649]

warning: Could not load shared library symbols for 3 libraries, e.g.
/lib/aarch64-linux-gnu/libtinfo.so.6.
Use the "info sharedlibrary" command to see the complete listing.
Do you need "set solib-search-path" or "set sysroot"?
Core was generated by `-bash'.
#0  0x0000ffffbafa6734 in ?? ()
```

8.   Verify that the stack trace (backtrace) is shown correctly with symbols:

```
(gdb) bt
#0  0x0000ffffbafa6734 in ?? ()
#1  0x0000000000000001 in ?? ()
Backtrace stopped: previous frame identical to this frame (corrupt stack?)
```

9.   If the stack trace on your system is incorrect, like in the output above, check that shared libraries are loaded, and if not, add the current directory to the shared library search path:

```
(gdb) info sharedlibrary
From                To                Syms Read    Shared Object Library
                                      No           /lib/aarch64-linux-gnu/libtinfo.so.6
                                      No           /lib/aarch64-linux-gnu/libc.so.6
                                      No           /lib/ld-linux-aarch64.so.1

(gdb) set solib-search-path .
Reading symbols from /home/coredump/LAPI/A64/libtinfo.so.6...(no debugging symbols
found)...done.
Reading symbols from /home/coredump/LAPI/A64/libc.so.6...(no debugging symbols found)...done.
Reading symbols from /home/coredump/LAPI/A64/ld-linux-aarch64.so.1...(no debugging symbols
found)...done.

(gdb) bt
#0  0x0000ffffbafa6734 in wait4 () from /home/coredump/LAPI/A64/libc.so.6
#1  0x0000aaaabb7a943c in ?? ()
#2  0x0000aaaabb70dd50 in wait_for ()
#3  0x0000aaaabb6f0dac in execute_command_internal ()
#4  0x0000aaaabb6f160c in execute_command ()
#5  0x0000aaaabb6e168c in reader_loop ()
#6  0x0000aaaabb6d4170 in main ()
```

10.   We exit GDB.

```
(gdb) q

~/LAPI/A64$
```

Exercise L0 (WinDbg)

Goal: Install WinDbg or Debugging Tools for Windows, or pull Docker image, and check that symbols are set up correctly.

Patterns: Stack Trace; Incorrect Stack Trace.

1.   Download memory dump files if you haven't done that already and unpack the archives:

https://www.patterndiagnostics.com/Training/LAPI/LAPI-Dumps.zip

2.   Install WinDbg (or upgrade existing WinDbg Preview) from https://learn.microsoft.com/en-gb/windows-hardware/drivers/debugger. Run WinDbg.

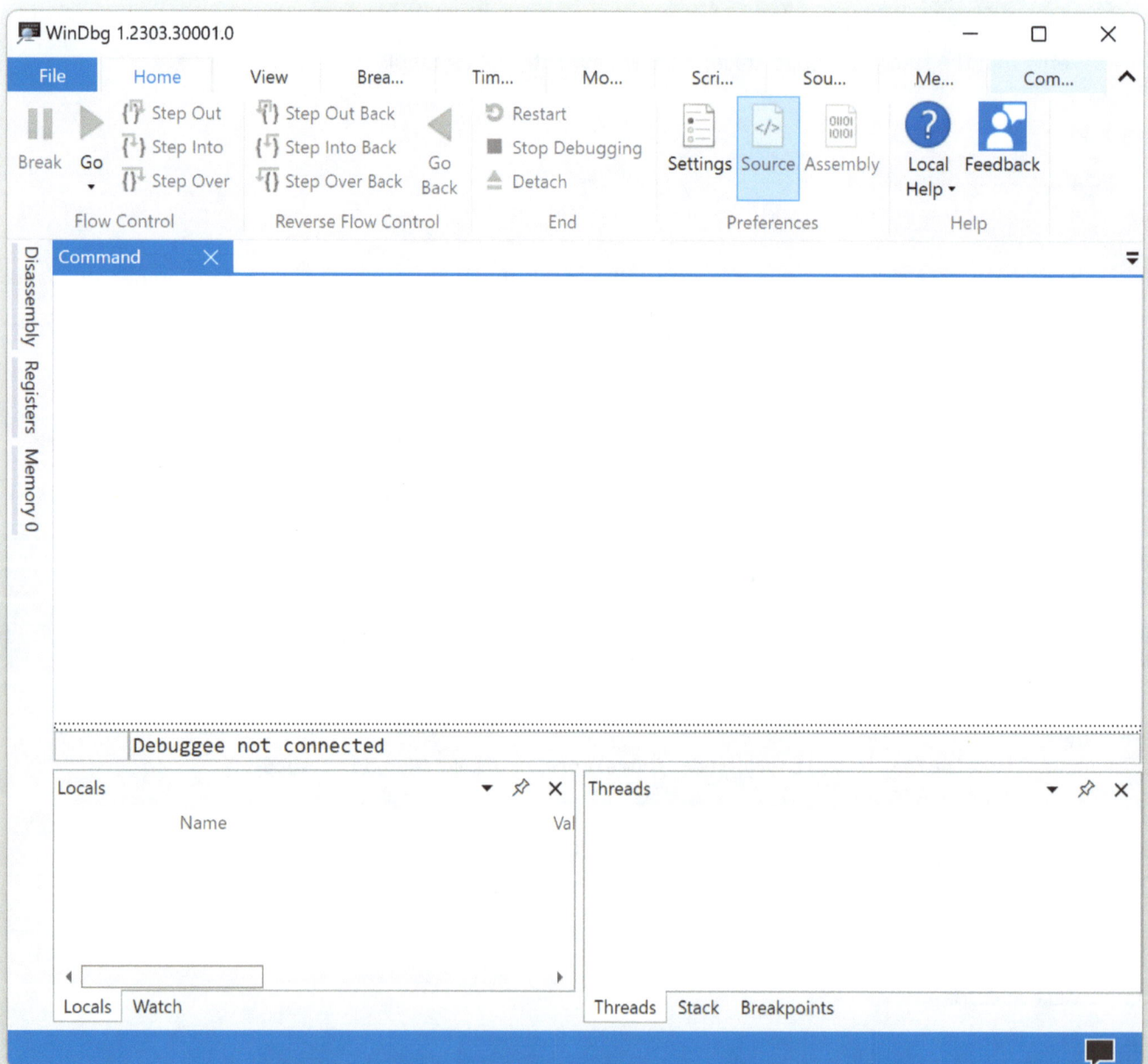

3.   Open \LAPI\A64\core.19649:

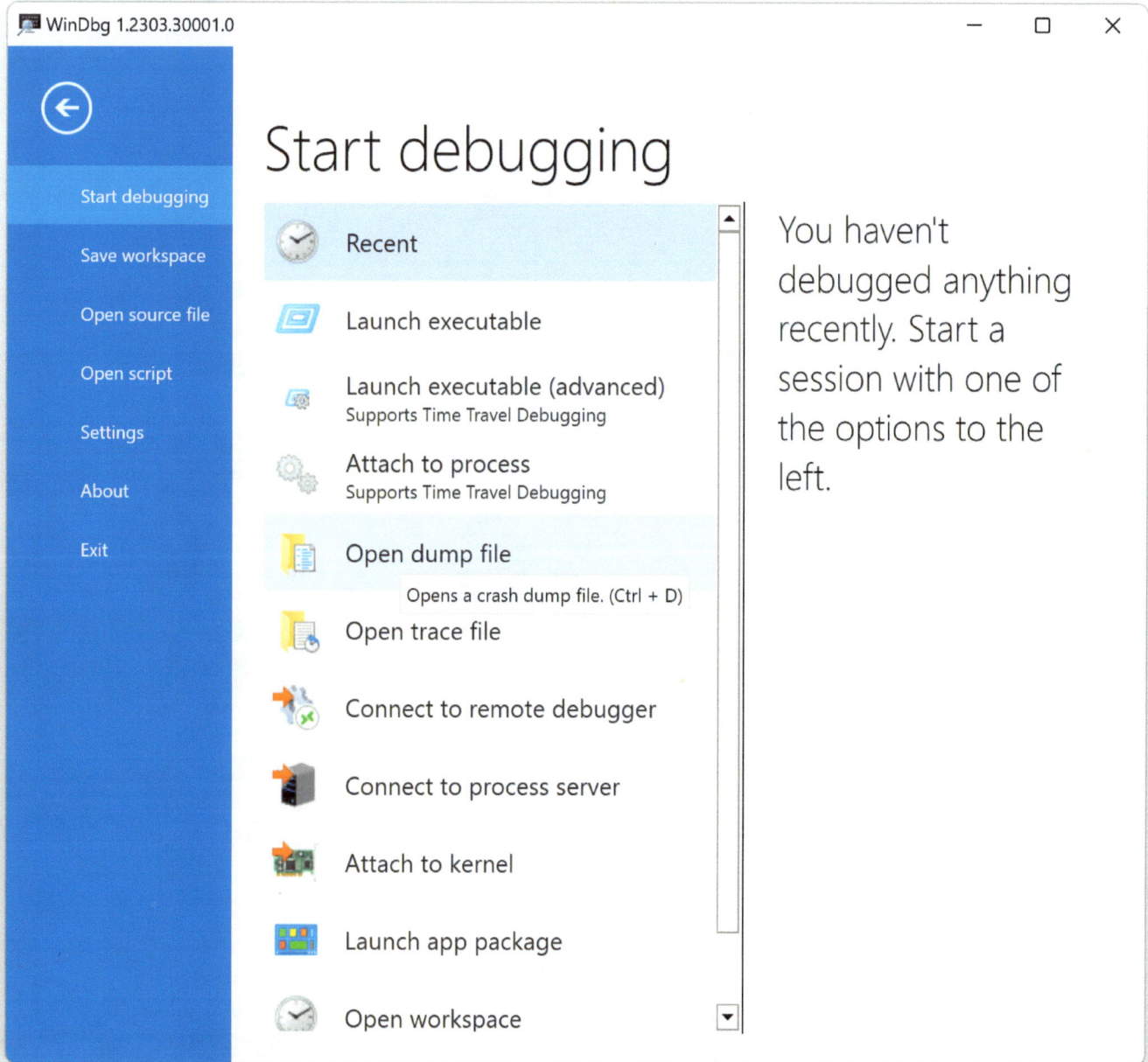

4.   We get the dump file loaded:

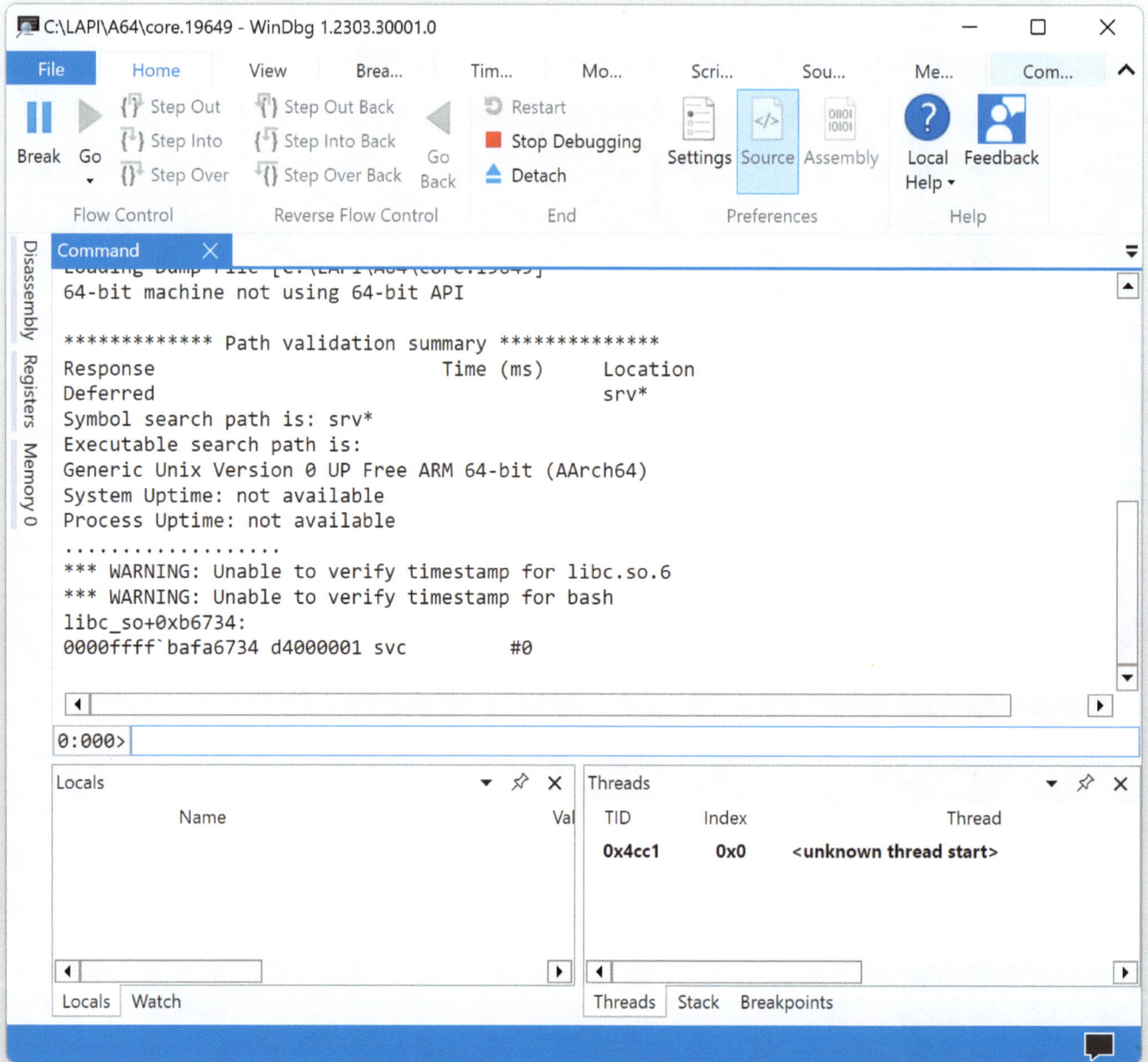

5.   Type **.sympath+** <path> command to set symbol path:

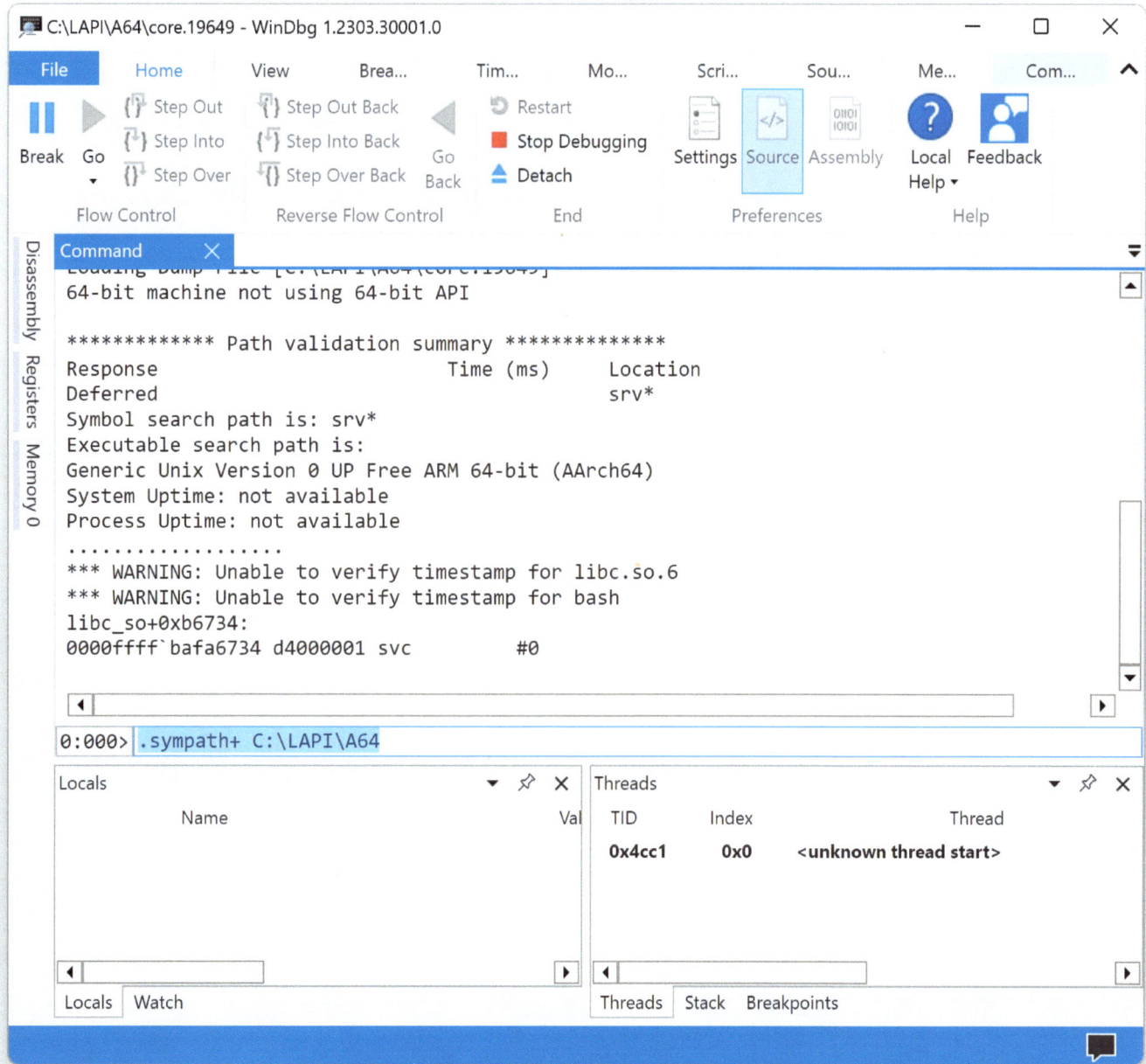

6.   Type **.reload** command to reload symbols:

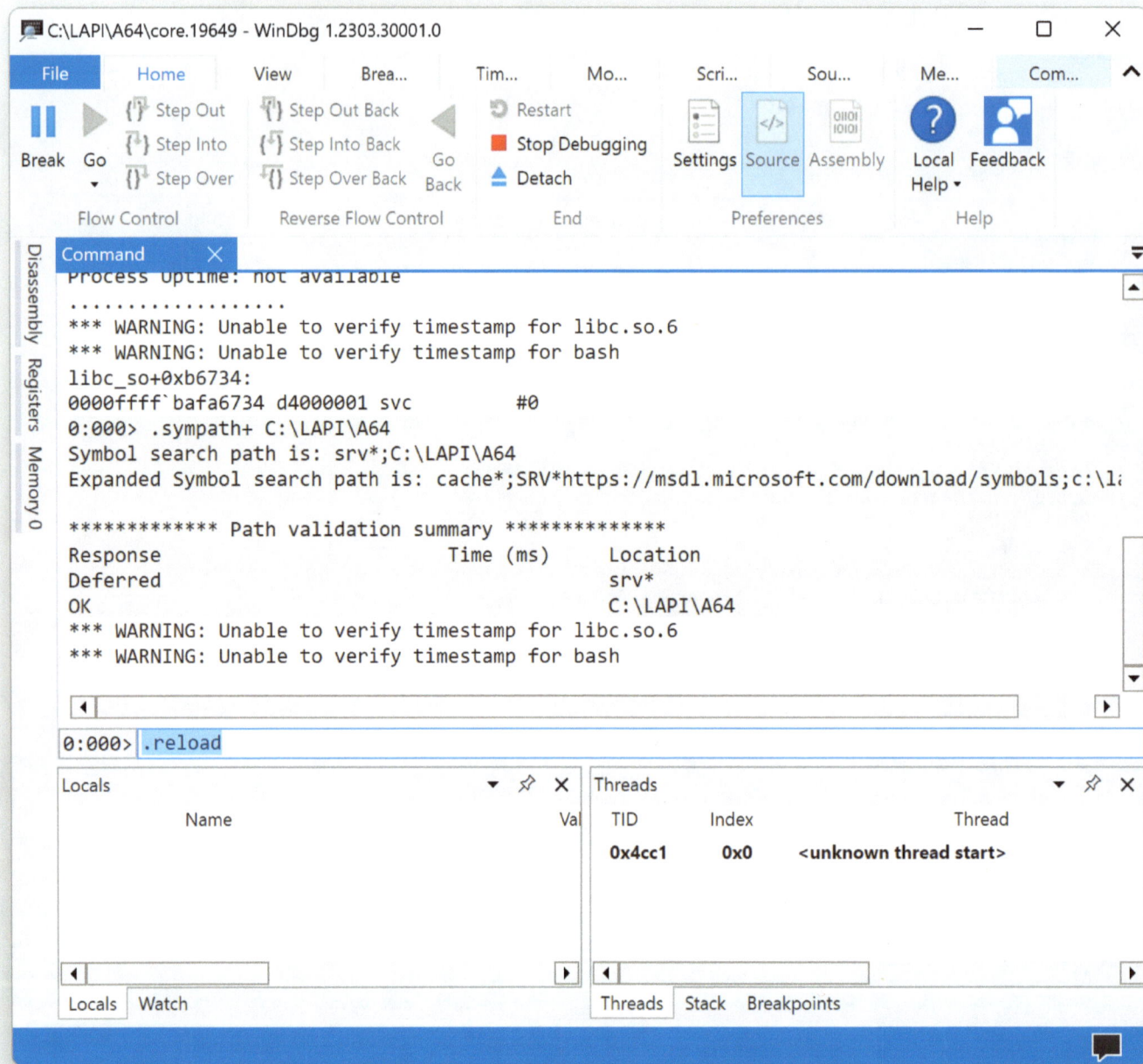

7.	Type **k** command to verify the correctness of the stack trace:

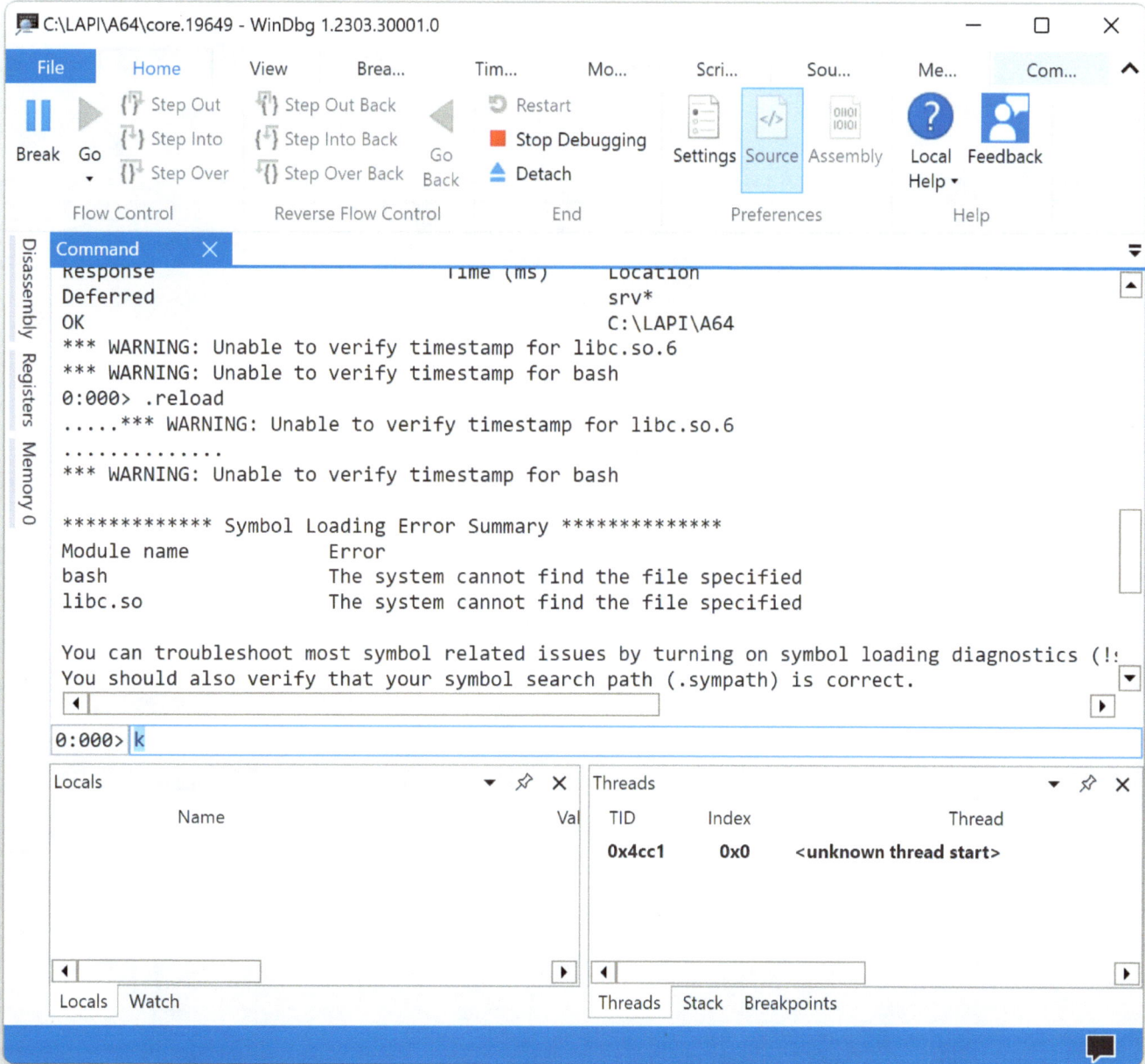

```
Response                Time (ms)       Location
Deferred                                srv*
OK                                      C:\LAPI\A64
*** WARNING: Unable to verify timestamp for libc.so.6
*** WARNING: Unable to verify timestamp for bash
0:000> .reload
.....*** WARNING: Unable to verify timestamp for libc.so.6
..............
*** WARNING: Unable to verify timestamp for bash

************* Symbol Loading Error Summary **************
Module name             Error
bash                    The system cannot find the file specified
libc.so                 The system cannot find the file specified

You can troubleshoot most symbol related issues by turning on symbol loading diagnostics (!
You should also verify that your symbol search path (.sympath) is correct.
```

```
0:000> k
```

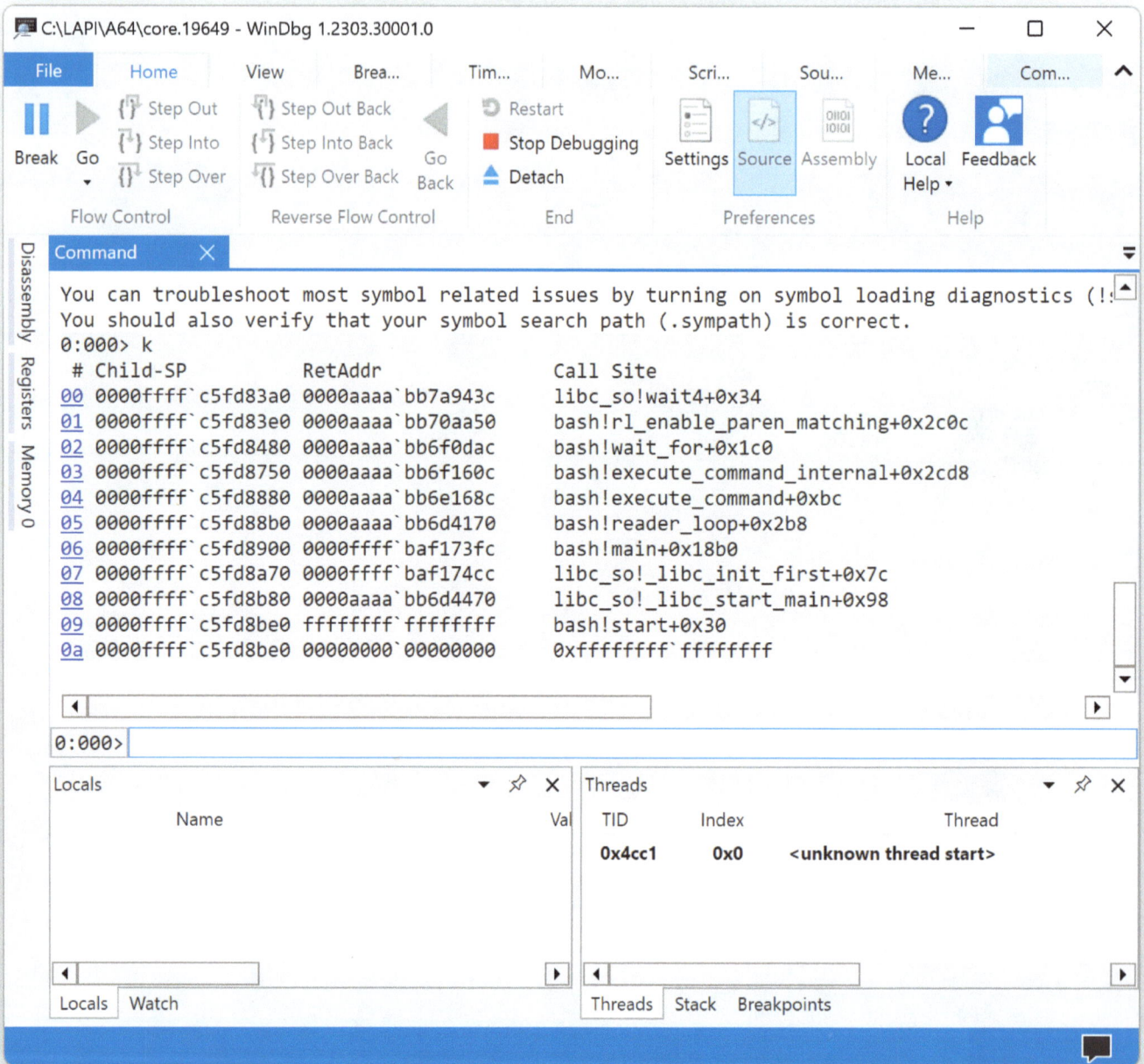

8.   The output of command should be this:

```
0:000> k
# Child-SP          RetAddr           Call Site
00 0000ffff`c5fd83a0 0000aaaa`bb7a943c  libc_so!wait4+0x34
01 0000ffff`c5fd83e0 0000aaaa`bb70aa50  bash!rl_enable_paren_matching+0x2c0c
02 0000ffff`c5fd8480 0000aaaa`bb6f0dac  bash!wait_for+0x1c0
03 0000ffff`c5fd8750 0000aaaa`bb6f160c  bash!execute_command_internal+0x2cd8
04 0000ffff`c5fd8880 0000aaaa`bb6e168c  bash!execute_command+0xbc
05 0000ffff`c5fd88b0 0000aaaa`bb6d4170  bash!reader_loop+0x2b8
06 0000ffff`c5fd8900 0000ffff`baf173fc  bash!main+0x18b0
07 0000ffff`c5fd8a70 0000ffff`baf174cc  libc_so!_libc_init_first+0x7c
08 0000ffff`c5fd8b80 0000aaaa`bb6d4470  libc_so!_libc_start_main+0x98
09 0000ffff`c5fd8be0 ffffffff`ffffffff  bash!start+0x30
0a 0000ffff`c5fd8be0 00000000`00000000  0xffffffff`ffffffff
```

If it has this form below with large offsets, then your symbol files were not set up correctly - **Incorrect Stack Trace** pattern:

```
0:000> k
# Child-SP          RetAddr            Call Site
00 0000ffff`c5fd83a0 0000aaaa`bb7a943c libc_so+0xb6734
01 0000ffff`c5fd83e0 0000aaaa`bb70aa50 bash!rl_enable_paren_matching+0x2c0c
02 0000ffff`c5fd8480 0000aaaa`bb6f0dac bash!wait_for+0x1c0
03 0000ffff`c5fd8750 0000aaaa`bb6f160c bash!execute_command_internal+0x2cd8
04 0000ffff`c5fd8880 0000aaaa`bb6e168c bash!execute_command+0xbc
05 0000ffff`c5fd88b0 0000aaaa`bb6d4170 bash!reader_loop+0x2b8
06 0000ffff`c5fd8900 0000ffff`baf173fc bash!main+0x18b0
07 0000ffff`c5fd8a70 0000ffff`baf174cc libc_so+0x273fc
08 0000ffff`c5fd8b80 0000aaaa`bb6d4470 libc_so+0x274cc
09 0000ffff`c5fd8be0 ffffffff`ffffffff bash!start+0x30
0a 0000ffff`c5fd8be0 00000000`00000000 0xffffffff`ffffffff
```

9.   [Optional] Download and install the latest version of Debugging Tools for Windows (See windbg.org for quick links, WinDbg Quick Links \ Download Debugging Tools for Windows). For this part, we use WinDbg 10.0.22621.382 from Windows 11 WDK, version 22H2.

10.   Launch WinDbg from Windows Kits \ WinDbg (X64) or Windows Kits \ WinDbg (X86). For uniformity, we use the X64 version of WinDbg throughout the exercises.

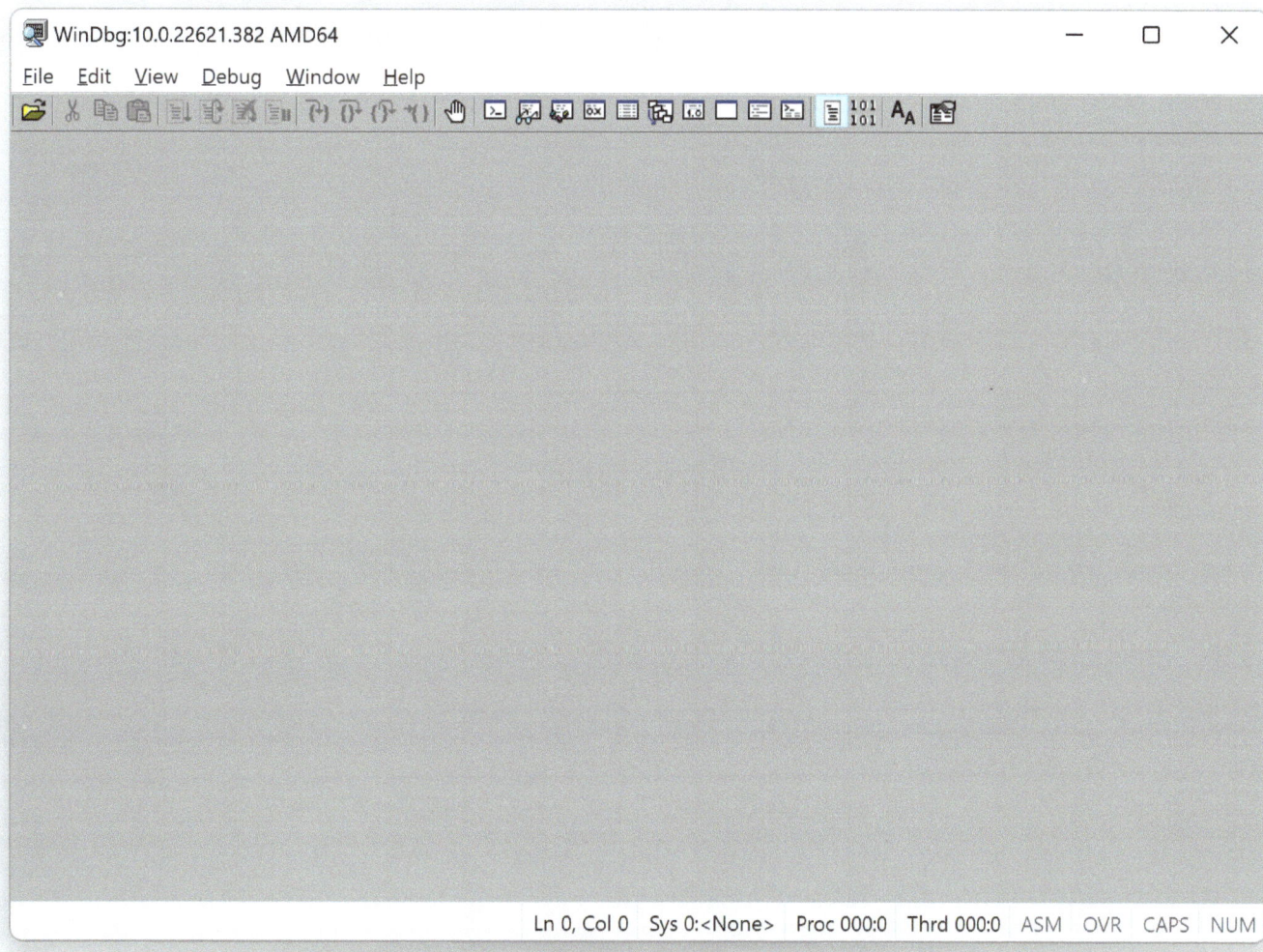

11. Open \LAPI\A64\core.19649:

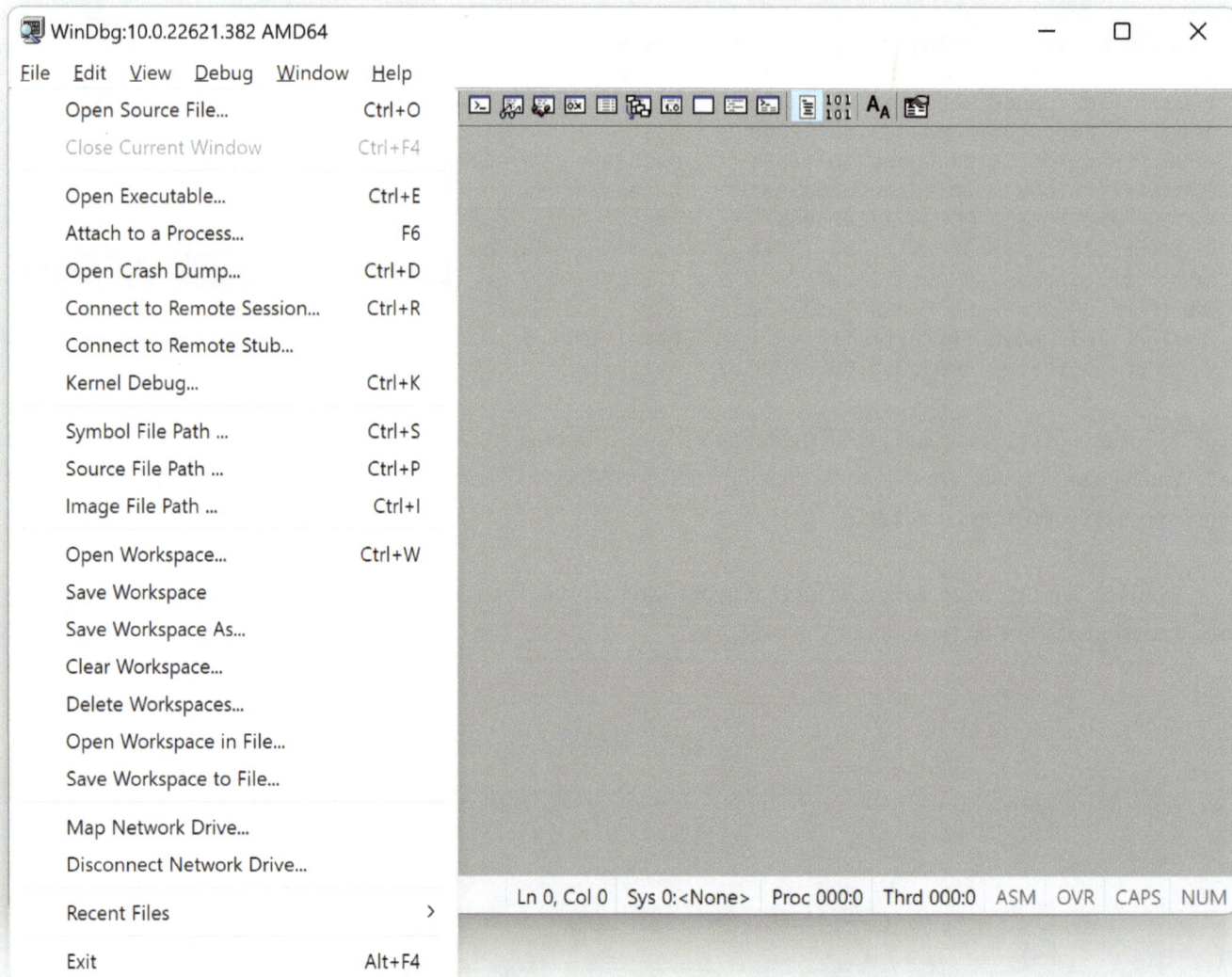

12.   We get the dump file loaded:

```
Dump C:\LAPI\A64\core.19649 - WinDbg:10.0.22621.382 AMD64

File  Edit  View  Debug  Window  Help

Command - Dump C:\LAPI\A64\core.19649 - WinDbg:10.0.22621.382 AMD64

Microsoft (R) Windows Debugger Version 10.0.22621.382 AMD64
Copyright (c) Microsoft Corporation. All rights reserved.


Loading Dump File [C:\LAPI\A64\core.19649]
64-bit machine not using 64-bit API

************* Path validation summary **************
Response                      Time (ms)      Location
Deferred                                     srv*
Symbol search path is: srv*
Executable search path is:
Generic Unix Version 0 UP Free ARM 64-bit (AArch64)
Machine Name:
System Uptime: not available
Process Uptime: not available
..................
*** WARNING: Unable to verify timestamp for libc.so.6
*** WARNING: Unable to verify timestamp for bash
libc_so+0xb6734:
0000ffff`bafa6734 d4000001 svc             #0

0:000>

Ln 0, Col 0   Sys 0:Target   Proc 000:4cc1   Thrd 000:4cc1   ASM  OVR  CAPS  NUM
```

33

13. Type **.sympath+** <path> command to set symbol path:

```
Command - Dump C:\LAPI\A64\core.19649 - WinDbg:10.0.22621.382 AMD64        —  □  ✕

Microsoft (R) Windows Debugger Version 10.0.22621.382 AMD64
Copyright (c) Microsoft Corporation. All rights reserved.


Loading Dump File [C:\LAPI\A64\core.19649]
64-bit machine not using 64-bit API

************* Path validation summary **************
Response                      Time (ms)      Location
Deferred                                     srv*
Symbol search path is: srv*
Executable search path is:
Generic Unix Version 0 UP Free ARM 64-bit (AArch64)
Machine Name:
System Uptime: not available
Process Uptime: not available
...................
*** WARNING: Unable to verify timestamp for libc.so.6
*** WARNING: Unable to verify timestamp for bash
libc_so+0xb6734:
0000ffff`bafa6734 d4000001 svc            #0

0:000> .sympath+ C:\LAPI\A64
```

14. Type **.reload** command to reload symbols:

```
Command - Dump C:\LAPI\A64\core.19649 - WinDbg:10.0.22621.382 AMD64        —  □  ✕
Response                      Time (ms)      Location
Deferred                                     srv*
Symbol search path is: srv*
Executable search path is:
Generic Unix Version 0 UP Free ARM 64-bit (AArch64)
Machine Name:
System Uptime: not available
Process Uptime: not available
...................
*** WARNING: Unable to verify timestamp for libc.so.6
*** WARNING: Unable to verify timestamp for bash
libc_so+0xb6734:
0000ffff`bafa6734 d4000001 svc        #0
0:000> .sympath+ C:\LAPI\A64
Symbol search path is: srv*;C:\LAPI\A64
Expanded Symbol search path is: cache*;SRV*https://msdl.microsoft.com/download/symbols;c:\lapi\a

************* Path validation summary **************
Response                      Time (ms)      Location
Deferred                                     srv*
OK                                           C:\LAPI\A64
*** WARNING: Unable to verify timestamp for libc.so.6


0:000> .reload
```

15.   Type the **k** command to verify the correctness of the stack trace:

```
Command - Dump C:\LAPI\A64\core.19649 - WinDbg:10.0.22621.382 AMD64          □    ✕
0000ffff`bafa6734 d4000001 svc            #0
0:000> .sympath+ C:\LAPI\A64
Symbol search path is: srv*;C:\LAPI\A64
Expanded Symbol search path is: cache*;SRV*https://msdl.microsoft.com/download/symbols;c:\lapi\a

************* Path validation summary **************
Response                     Time (ms)      Location
Deferred                                    srv*
OK                                          C:\LAPI\A64
*** WARNING: Unable to verify timestamp for libc.so.6
0:000> .reload
.................
*** WARNING: Unable to verify timestamp for libc.so.6
*** WARNING: Unable to verify timestamp for bash

************* Symbol Loading Error Summary **************
Module name          Error
bash                 The system cannot find the file specified
libc.so              The system cannot find the file specified

You can troubleshoot most symbol related issues by turning on symbol loading diagnostics (!sym n
You should also verify that your symbol search path (.sympath) is correct.

 ████████████████████████

0:000> k
```

```
Command - Dump C:\LAPI\A64\core.19649 - WinDbg:10.0.22621.382 AMD64          □    ✕
*** WARNING: Unable to verify timestamp for bash

************* Symbol Loading Error Summary **************
Module name          Error
bash                 The system cannot find the file specified
libc.so              The system cannot find the file specified

You can troubleshoot most symbol related issues by turning on symbol loading diagnostics (!sym n
You should also verify that your symbol search path (.sympath) is correct.
0:000> k
 # Child-SP          RetAddr               Call Site
00 0000ffff`c5fd83a0 0000aaaa`bb7a943c    libc_so!wait4+0x34
01 0000ffff`c5fd83e0 0000aaaa`bb70aa50    bash!rl_enable_paren_matching+0x2c0c
02 0000ffff`c5fd8480 0000aaaa`bb6f0dac    bash!wait_for+0x1c0
03 0000ffff`c5fd8750 0000aaaa`bb6f160c    bash!execute_command_internal+0x2cd8
04 0000ffff`c5fd8880 0000aaaa`bb6e168c    bash!execute_command+0xbc
05 0000ffff`c5fd88b0 0000aaaa`bb6d4170    bash!reader_loop+0x2b8
06 0000ffff`c5fd8900 0000ffff`baf173fc    bash!main+0x18b0
07 0000ffff`c5fd8a70 0000ffff`baf174cc    libc_so!_libc_init_first+0x7c
08 0000ffff`c5fd8b80 0000aaaa`bb6d4470    libc_so!_libc_start_main+0x98
09 0000ffff`c5fd8be0 ffffffff`ffffffff    bash!start+0x30
0a 0000ffff`c5fd8be0 00000000`00000000    0xffffffff`ffffffff

 ████████████████████████

0:000>
```

16.   [Optional] If you prefer using a Docker image with WinDbg and symbol files included, follow these steps below.

```
c:\LAPI>docker pull patterndiagnostics/windbg:10.0.22621.382-lapi
10.0.22621.382-lapi: Pulling from patterndiagnostics/windbg
Digest: sha256:e4ccb0b18a3dacd70e7ded51ea3e6e3c15bad6b8428ed7094fe214ce0262d98e
Status: Image is up to date for patterndiagnostics/windbg:10.0.22621.382-lapi
docker.io/patterndiagnostics/windbg:10.0.22621.382-lapi


c:\LAPI>docker run -it -v C:\LAPI:C:\LAPI patterndiagnostics/windbg:10.0.22621.382-lapi
Microsoft Windows [Version 10.0.20348.1726]
(c) Microsoft Corporation. All rights reserved.


C:\WinDbg>windbg.bat C:\LAPI\A64\core.19649


Microsoft (R) Windows Debugger Version 10.0.22621.382 AMD64
Copyright (c) Microsoft Corporation. All rights reserved.



Loading Dump File [C:\LAPI\A64\core.19649]
64-bit machine not using 64-bit API


************* Path validation summary **************
Response                        Time (ms)       Location
OK                                                  .
Symbol search path is: .
Executable search path is:
Generic Unix Version 0 UP Free ARM 64-bit (AArch64)
Machine Name:
System Uptime: not available
Process Uptime: not available
...................
*** WARNING: Unable to verify timestamp for libc.so.6
*** WARNING: Unable to verify timestamp for bash
libc_so+0xb6734:
0000ffff`bafa6734 d4000001 svc         #0


0:000> .sympath+ C:\LAPI\A64
Symbol search path is: .;C:\LAPI\A64
Expanded Symbol search path is: .;c:\lapi\a64


************* Path validation summary **************
Response                        Time (ms)       Location
OK                                                  .
OK                                              C:\LAPI\A64


0:000> .reload
...................
*** WARNING: Unable to verify timestamp for libc.so.6
*** WARNING: Unable to verify timestamp for bash


************* Symbol Loading Error Summary **************
Module name          Error
bash                 The system cannot find the file specified
libc.so              The system cannot find the file specified


You can troubleshoot most symbol related issues by turning on symbol loading diagnostics (!sym
noisy) and repeating the command that caused symbols to be loaded.
You should also verify that your symbol search path (.sympath) is correct.
```

```
0:000> k
Child-SP          RetAddr           Call Site
0000ffff`c5fd83a0 0000aaaa`bb7a943c libc_so!wait4+0x34
0000ffff`c5fd83e0 0000aaaa`bb70aa50 bash!rl_enable_paren_matching+0x2c0c
0000ffff`c5fd8480 0000aaaa`bb6f0dac bash!wait_for+0x1c0
0000ffff`c5fd8750 0000aaaa`bb6f160c bash!execute_command_internal+0x2cd8
0000ffff`c5fd8880 0000aaaa`bb6e168c bash!execute_command+0xbc
0000ffff`c5fd88b0 0000aaaa`bb6d4170 bash!reader_loop+0x2b8
0000ffff`c5fd8900 0000ffff`baf173fc bash!main+0x18b0
0000ffff`c5fd8a70 0000ffff`baf174cc libc_so!_libc_init_first+0x7c
0000ffff`c5fd8b80 0000aaaa`bb6d4470 libc_so!_libc_start_main+0x98
0000ffff`c5fd8be0 ffffffff`ffffffff bash!start+0x30
0000ffff`c5fd8be0 00000000`00000000 0xffffffff`ffffffff

0:000> q
quit:
NatVis script unloaded from 'C:\Program Files\Windows Kits\10\Debuggers\x64\Visualizers\
gstl.natvis'

C:\WinDbg>exit

c:\LAPI>
```

Why Linux API?

- Development
- Malware analysis
- Vulnerability analysis and exploitation
- Reversing
- Diagnostics
- Debugging
- Monitoring
- Memory forensics
- Crash and hang analysis
- Secure coding
- Static code analysis
- Trace and log analysis

First, why did we create this course? The knowledge of Linux API is necessary for many software construction and post-construction activities listed on this slide. In this training, we look at Linux API from a software diagnostics perspective. This perspective includes core memory dump analysis and, partially, trace and log analysis. The knowledge of Linux API is tacitly assumed in my other courses. Course attendees sometimes ask why I chose a particular function from a backtrace for analysis or what this or that function is doing. Of course, there is an intersection of what we learn with other areas as well. During this course, we also do a bit of live debugging.

My History of Linux API

- ◉ C runtime library from 1989
- ◉ I started using Linux API in 1997 (Old CV)
- ◉ Commercial application development in 2000 – 2003
- ◉ Reading of Linux API for core dump analysis since 2015
- ◉ Teaching foundations of x64 Linux debugging from 2021
- ◉ and A64 Linux debugging from 2022, two books
- ◉ System programming using Linux API in 2022
- ◉ Linux disassembly and reversing course in 2022,
- ◉ second edition in 2023
- ◉ Third edition of Linux core dump analysis course in 2023
- ◉ Commercial core dump analysis service in 2023

It is hard for me to recall when I started using Linux API. It is a distant memory but not as distant as Windows API. I believe it was the mid-1990s when I at least installed Linux a few times. Fortunately, I recalled that I meticulously documented my work history before 2003 in my old CV, which I no longer maintain. So, from the record, I see I started using Linux API in 1997. In 2000, I designed and partly implemented a multiplatform application for Windows, Linux, and Solaris. Linux was my primary development system for C++ static code analysis tools in 2001 – 2003, with Emacs as the code editor and GNU tool chain for development. Then in mid-2003, I switched completely to Windows development. In 2012 I started working on macOS (at that time called Mac OS X) core dump analysis after a brief encounter with FreeBSD internals and published the course and its second edition in 2014 with added LLDB material. Then in 2015, I started doing Linux core dump analysis and published the first edition of the course. In 2017 – 2020, I studied Linux internals while preparing for various interviews. Finally, in mid-2020, I switched to full-time Linux-based cloud development. Throughout 2022, I used Linux API daily to implement Linux API monitoring tools. In 2023 I continued updating various Linux courses and even started commercial Linux core dump analysis with the success of diagnosing customer issues and providing recommendations.

Old CV
https://opentask.com/Vostokov/CV.htm

Perspectives of Linux API

◉ Memory analysis: dumps / live debugging

◉ Disassembly, reconstruction, reversing

◉ Trace and log analysis (strace, ltrace, perf, eBPF, procmon)

◉ Category theory

The perspective we take is from software diagnostics, particularly memory analysis (this may include memory forensics), which includes both dump analysis and live debugging, disassembly, reconstruction, reversing (the so-called ADDR patterns), and trace and log analysis using tools such as strace, ltrace, perf, eBPF, and even procmon for Linux. We do not look at Linux API from a development perspective, such as writing applications and services, although there is some necessary intersection, especially when doing diagnostics and vulnerability analysis. We also take a view of Linux API from a category theory perspective and compare it to Windows API as well.

Procmon
https://github.com/Sysinternals/ProcMon-for-Linux

What Linux API?

◉ Source code perspective

◉ ABI (Application Binary Interface) perspective

- Libraries
- Syscalls

Let's determine what we mean by Linux API. There are two general perspectives on this question. The first is the source code (or software construction) perspective. In this training, we take a different software post-construction perspective: what is seen from already compiled and linked modules is what we call Linux API.

API and Language Levels

- ◉ Conceptual cross-platform level

- ◉ High-level and assembly language

- ◉ Machine language

Conceptually, there are many similarities between different APIs, such as Windows and Linux. This is understandable since a large part of API centers on interfacing with OS, which must provide common resource abstractions. The next level is the high-level and assembly languages level, which differs across platforms. And finally, the machine language level erases all differences in the previous level.

x64 and A64

x64 and A64

In this section, we provide an overview of disassembly for x64 and ARM64 platforms. Windows developers who know x64 assembly language may benefit because we use a different flavor, default in Linux GDB. Compared to other Linux courses where I added the same review, it has slightly less material since we are not interested in stack trace reconstruction, for example. However, I added a small number of additional examples we see in this course's core dumps.

CPU Registers (x64)

- RAX ⊃ EAX ⊃ AX ⊇ {AH, AL} | RAX 64-bit | EAX 32-bit |

- ALU: **RAX**, **RDX**

- Counter: **RCX**

- Memory copy: **RSI** (src), **RDI** (dst)

- Stack: **RSP**, **RBP**

- Next instruction: **RIP**

- New: **R8 – R15**, **Rx(D|W|L)**

There are familiar 32-bit CPU register names, such as **EAX,** that are extended to 64-bit names, such as **RAX**. Most of them are traditionally specialized, such as ALU, counter, and memory copy registers. Although, now they all can be used as general-purpose registers. There is, of course, a stack pointer, **RSP**, and, additionally, a frame pointer, **RBP**, that is used to address local variables and saved parameters. It can be used for backtrace reconstruction. In some compiler code generation implementations, **RBP** is also used as a general-purpose register, with **RSP** taking the role of a frame pointer. An instruction pointer RIP is saved in the stack memory region with every function call, then restored on return from the called function. In addition, the x64 platform features another eight general-purpose registers, from **R8** to **R15**.

Instructions: registers (x64)

⊙ Opcode SRC, DST # default AT&T flavour

⊙ Examples:

```
mov    $0x10, %rax          # 0x10 → RAX
mov    %rsp, %rbp           # RSP → RBP
add    $0x10, %r10          # R10 + 0x10 → R10
imul   %ecx, %edx           # ECX * EDX → EDX
callq  *%rdx                # RDX already contains
                           #     the address of func (&func)
                           # PUSH RIP; &func → RIP
sub    $0x30, %rsp          # RSP-0x30 → RSP
                           # make a room for local variables
```

© 2023 Software Diagnostics Services

This slide shows a few examples of CPU instructions involving operations with registers, such as moving a value and doing arithmetic. The direction of operands is opposite to the Intel x64 disassembly flavor if you are accustomed to WinDbg on Windows. It is possible to use the Intel disassembly flavor in GDB, but we opted for the default AT&T flavor in line with our **Accelerated Linux Core Dump Analysis** book.

Memory and Stack Addressing

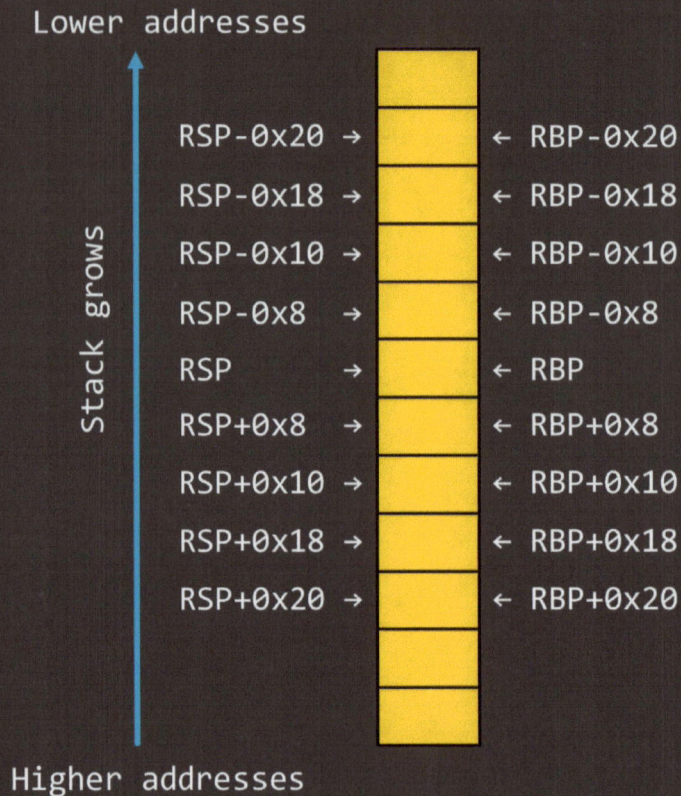

```
Lower addresses
   ↑
   |        RSP-0x20 →  [   ]  ← RBP-0x20
   |        RSP-0x18 →  [   ]  ← RBP-0x18
   |        RSP-0x10 →  [   ]  ← RBP-0x10
 Stack      RSP-0x8  →  [   ]  ← RBP-0x8
 grows      RSP      →  [   ]  ← RBP
   |        RSP+0x8  →  [   ]  ← RBP+0x8
   |        RSP+0x10 →  [   ]  ← RBP+0x10
   |        RSP+0x18 →  [   ]  ← RBP+0x18
   |        RSP+0x20 →  [   ]  ← RBP+0x20
   |                    [   ]
   |                    [   ]
Higher addresses
```

Before we look at operations with memory, let's look at a graphical representation of memory addressing. A thread stack is just any other memory region, so instead of **RSP** and **RBP,** any other register can be used. Please note that stack grows towards lower addresses, so to access the previously pushed values, you need to use positive offsets from **RSP**.

Instructions: memory load (x64)

- Opcode Offset(SRC), DST

- Opcode DST

- Examples:

```
mov    0x10(%rsp), %rax        # value at address RSP+0x10 → RAX
mov    -0x10(%rbp), %rcx       # value at address RBP-0x10 → RCX
add    (%rax), %rdx            # RDX + value at address RAX → RDX
pop    %rdi                    # value at address RSP → RDI
                              # RSP + 8 → RSP
lea    0x20(%rbp), %r8         # address RBP+0x20 → R8
```

Constants are encoded in instructions, but if we need arbitrary values, we must get them from memory. Round brackets show memory access relative to an address stored in some register.

Instructions: memory store (x64)

- Opcode SRC, Offset(DST)

- Opcode SRC|DST

- Examples:

```
mov    %rcx, -0x20(%rbp)      # RCX → value at address RBP-0x20
addl   $1, (%rax)             # 1 + 32-bit value at address RAX →
                              #     32-bit value at address RAX
push   %rsi                   # RSP - 8 → RSP
                              # RSI → value at address RSP
inc    (%rcx)                 # 1 + value at address RCX →
                              #     value at address RCX
```

Storing is similar to loading.

Instructions: flow (x64)

⊙ Opcode DST

⊙ Examples:

```
jmpq    0x10493fc1c        # 0x10493fc1c → RIP
                           # (goto 0x10493fc1c)


jmpq    *0x100(%rip)       # value at address RIP+0x100 → RIP


callq   0x10493ff74        # RSP – 8 → RSP
0x10493fc14:               # 0x10493fc14 → value at address RSP
                           # 0x10493ff74 → RIP
                           # (goto 0x10493ff74)
```

Goto (an unconditional jump) is implemented via the **JMP** instruction. Function calls are implemented via **CALL** instruction. For conditional branches, please look at the official Intel documentation. We don't use these instructions in our exercises.

Function Call and Prolog (x64)

```
# void proc(int p1, long p2);
mov  $0x1, %edi
mov  $0x2, %rsi
call proc
addr:

# void proc2();
# void proc(int p1, long p2) {
#    long local = 0;
#    proc2();
# }
proc:
push %rbp
mov  %rsp, %rbp
sub  $0x8, %rsp
mov  $0, -0x8(%rbp)
call proc2
adr2:
...
```

Lower addresses

Stack grows

RSP-0x20 →		← RBP-0x28
RSP →	adr2	← RBP-0x20
RSP-0x10 →	0	← RBP-0x8
RSP-0x8 →	RBP	← RBP
RSP →	addr	← RBP-0x8
RSP+0x8 →		← RBP
RSP+0x10 →		← RBP+0x8
RSP+0x18 →		← RBP+0x10
RSP+0x20 →		← RBP+0x18

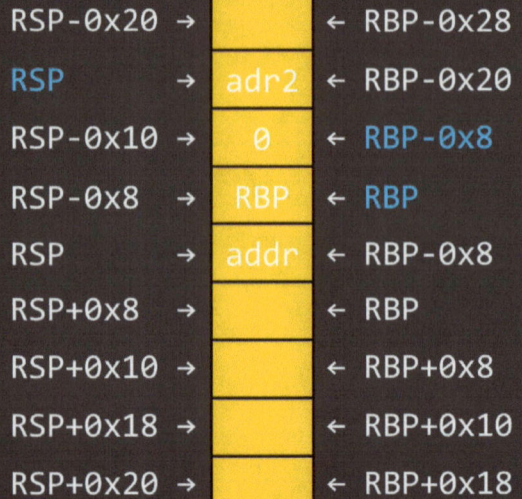

Higher addresses

When a function is called from the caller, a callee needs to do certain operations to make room for local variables on the thread stack. There are different ways to do that, and the assembly language code on the left is one of them. I use a different color in the diagram on the right to highlight the updated **RSP** and **RBP** values from the start of the *proc* function up to the moment when the *proc2* function is called. For simplicity of illustration, I only use 64-bit values.

CPU Registers (A64)

- **X0 – X28**, **W0 – W28**

X 64-bit	W 32-bit

- **X16** (**XIP0**), **X17** (**XIP1**)

- Stack: **SP**, **X29** (**FP**)

- Next instruction: **PC**

- Link register: **X30** (**LR**)

- Zero register: **XZR**, **WZR**

- 64-bit floating point registers **D0 – D31**

- 128-bit **Q0 – Q31**

There are 31 general registers from **X0** and **X30**, with some delegated to specific tasks such as intra-procedure calls (**X16**, **XIP0**, and **X17**, **XIP1**), addressing stack frames (Frame Pointer, **FP**, **X29**) and return addresses, the so-called Link Register (**LR**, **X30**). When you call a function, the return address of a caller is saved in **LR**, not on the stack as in Intel/AMD x64. The return instruction in a callee uses the address in **LR** to assign it to **PC** and resume execution. But if a callee calls other functions, the current **LR** needs to be manually saved somewhere, usually on the stack. There's Stack Pointer, **SP**, of course. To get zero values, there's the so-called Zero Register, **XZR**. All **X** registers are 64-bit, and 32-bit lower parts are addressed via the **W** prefix. There are also 128-bit SIMD registers. Next, we briefly look at some aspects related to our exercises.

Instructions: registers (A64)

- Opcode DST, SRC, SRC$_2$

- Examples:

```
mov   x0, #16           // X0 ← 16 (0x10)
mov   x29, sp           // X29 ← SP
add   x1, x2, #16       // X1 ← X2+16 (0x10)
mul   x1, x2, x3        // X1 ← X2*X3
blr   x8               // X8 already contains
                       //     the address of func (&func)
                       // LR ← PC+4; PC ← &func
sub   sp, sp, #48       // SP ← SP-48 (-0x30)
                       // make a room for local variables
```

This slide shows a few examples of CPU instructions that involve operations with registers, for example, moving a value and doing arithmetic. The direction of operands is the same as in the Intel x64 disassembly flavor if you are accustomed to WinDbg on Windows. It is equivalent to an assignment. **BLR** is a call of some function whose address is in the register. **BL** means Branch and Link.

Memory and Stack Addressing

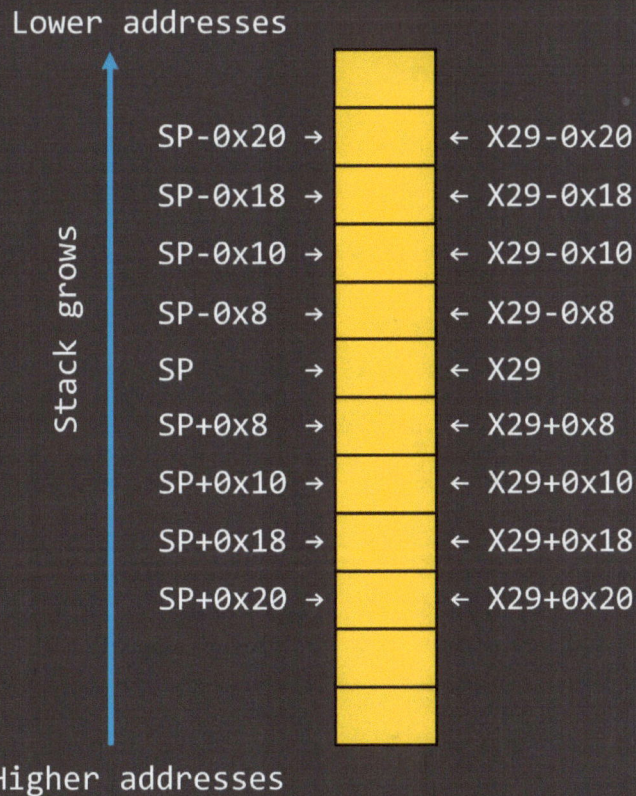

Lower addresses

SP-0x20 →		← X29-0x20
SP-0x18 →		← X29-0x18
SP-0x10 →		← X29-0x10
SP-0x8 →		← X29-0x8
SP →		← X29
SP+0x8 →		← X29+0x8
SP+0x10 →		← X29+0x10
SP+0x18 →		← X29+0x18
SP+0x20 →		← X29+0x20

Stack grows

Higher addresses

Before we look at operations with memory, let's look at a graphical representation of memory addressing. A thread stack is just any other memory region, so instead of **SP** and **X29 (FP)**, any other register can be used. Please note that the stack grows towards lower addresses, so to access the previously pushed values, you need to use positive offsets from **SP**.

Instructions: memory load (A64)

- Opcode DST, DST$_2$, [SRC, Offset]

- Opcode DST, DST$_2$, [SRC], Offset // Postincrement

- Examples:

```
ldr    x0, [sp]              // X0 ← value at address SP+0
ldr    x0, [x29, #-8]        // X0 ← value at address X29-0x8
ldp    x29, x30, [sp, #32]   // X29 ← value at address SP+32 (0x20)
                             // X30 ← value at address SP+40 (0x28)
ldp    x29, x30, [sp], #16   // X29 ← value at address SP+0
                             // X30 ← value at address SP+8
                             // SP ← SP+16 (0x10)
```

Constants are encoded in instructions, but if we need arbitrary values, we must get them from memory. Square brackets are used to show memory access relative to an address stored in some register. There's also an option to adjust the value of the register after load, the so-called **Postincrement**, which can be negative. As we see later, loading pairs of registers can be useful.

Instructions: memory store (A64)

- Opcode SRC, SRC$_2$, [DST, Offset]

- Opcode SRC, SRC$_2$, [DST, Offset]! // Preincrement

- Examples:

```
str    x0, [sp, #16]          // x0 → value at address SP+16 (0x10)
str    x0, [x29, #-8]         // x0 → value at address X29-8
stp    x29, x30, [sp, #32]    // x29 → value at address SP+32 (0x20)
                              // x30 → value at address SP+40 (0x28)
stp    x29, x30, [sp, #-16]!  // SP ← SP-16 (-0x10)
                              // x29 → set value at address SP
                              // x30 → set value at address SP+8
```

Storing operand order goes in the other direction compared to other instructions. There's a possibility to **Preincrement** the destination register before storing values.

Instructions: flow (A64)

- Opcode **DST**, SRC

- Examples:

```
adrp  x0, 0x420000      // x0 ← 0x420000

b       0x10493fc1c     // PC ← 0x10493fc1c
                        // (goto 0x10493fc1c)
br      x17             // PC ← the value of X17


0x10493fc14:            // PC == 0x10493fc14
bl      0x10493ff74     // LR ← PC+4 (0x10493fc18)
                        // PC ← 0x10493ff74
                        // (goto 0x10493ff74)
```

Because the size of every instruction is 4 bytes (32 bits), it is only possible to encode a part of a large 4GB address range, either as a relative offset to the current **PC** or via **ADRP** instruction. Goto (an unconditional branch) is implemented via the **B** instruction. Function calls are implemented via the **BL** (Branch and Link) instruction.

Function Call and Prolog (A64)

Lower addresses

X30 adr2

```
// void proc(int p1, long p2);
mov  w0, #0x1
mov  x1, #0x2
bl   proc
addr:

// void proc2();
// void proc(int p1, long p2) {
//   long local = 0;
//   proc2();
// }
proc:
stp  x29, x30, [sp, #-32]!
mov  x29, sp
str  zxr, [x29, #16]
bl   proc2
adr2:
...
```

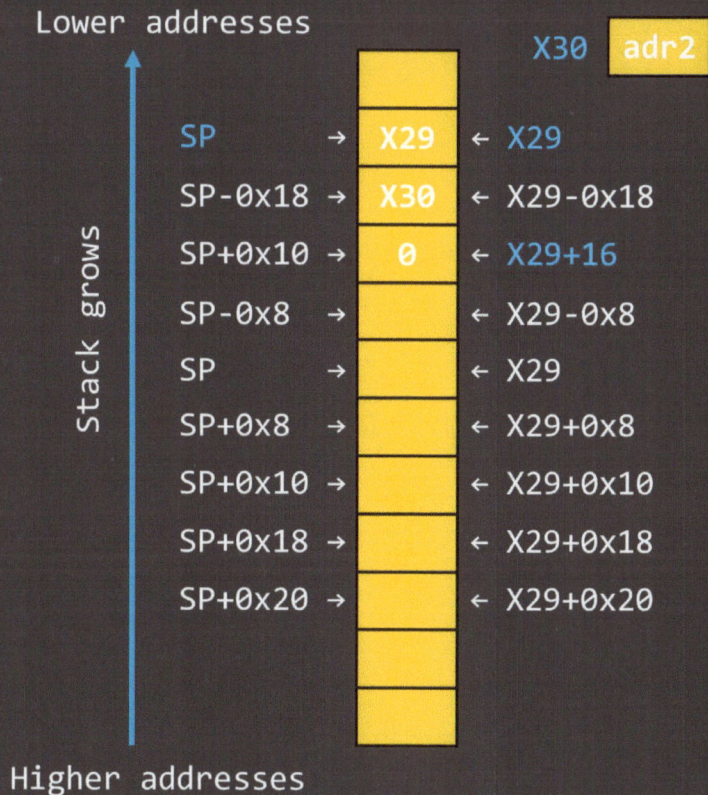

Stack grows

SP →	X29	← X29
SP-0x18 →	X30	← X29-0x18
SP+0x10 →	0	← X29+16
SP-0x8 →		← X29-0x8
SP →		← X29
SP+0x8 →		← X29+0x8
SP+0x10 →		← X29+0x10
SP+0x18 →		← X29+0x18
SP+0x20 →		← X29+0x20

Higher addresses

© 2023 Software Diagnostics Services

When a function is called from the caller, a callee needs to do certain operations to make room for local variables on the thread stack and save **LR** if there are further calls in the function body. There are different ways to do that, and the assembly language code on the left is one of them GCC compiler uses by default. I use a different color in the diagram on the right to highlight the updated **SP** and **X29** (**FP**) values from the start of the *proc* function up to the moment when the *proc2* function is called. Please also note an example of zero register usage and the fact that both **SP** and **X29** point to the same memory location. Positive offsets from **X29** are used to address local variables. For simplicity of illustration, I only use 64-bit values.

General Linux API Aspects

General Linux API Aspects

Lexicon

- IAT ↔ PLT
- DLL ↔ SO
- Module ↔ Shared Library

On this slide, I keep adding some vocabulary that differs between the standard accounts of Windows and Linux APIs but refers to the same concepts. IAT is Import Address Table, PLT is Procedure Linkage Table.

Manual Page and Header Views

- ⊙ Manual pages (2 and 3)

  - Syscalls / overview
  - Library calls

- ⊙ Headers

  - Kernel source code cross-reference
  - Manual pages 0

If you are unfamiliar with C or C++, a header is a textual file referenced in source code and inserted during compilation. It can contain anything but most likely a programming language source code such as function declarations, in our context, for example, declarations of Linux API functions and types. So, for example, if we are interested in creating threads, the *pthread_create* API function is declared in the *pthread.h* header file.

Syscalls
https://man7.org/linux/man-pages/dir_section_2.html

Overview
https://man7.org/linux/man-pages/man2/syscalls.2.html

Kernel source code cross-reference
https://elixir.bootlin.com/linux/latest/source

Manual page 0
https://man7.org/linux/man-pages/man0/

Naming Convention

- Naming conventions

- Functions, types, parameters, fields: twowords, snake_case

  - pthread_create
  - pid_t
  - tm_sec

- Constants, macros: TWOWORDS, SCREAMING_SNAKE_CASE

  - SOCK_STREAM

There are several naming conventions. The Linux API convention differs significantly from the Windows API convention; also, there are differences between different programming languages; please see the first link on the slide. Linux API naming connection is close to Standard C and C++ library naming conventions. Linux API uses **twowords** and **snake_case** for functions, types, parameters, and structure fields. Constants and macros use a different snake case convention called **SCREAMING_SNAKE_CASE** and also **TWOWORDS**.

Naming conventions
https://en.wikipedia.org/wiki/Naming_convention_(programming)

Snake Case
https://en.wikipedia.org/wiki/Snake_case

Basic Type System

- ◎ Basic types are `typedef-ed`

- ◎ `inttypes.h`

  - • `uint64_t`

- ◎ `sys/types.h`

  - • `size_t, pid_t`

GDB Commands
`(gdb) info types`
`(gdb) info functions`

WinDbg Commands
`0:000> dt *!*`
`0:000> x *!*`

Linux API uses a C language type system with additional custom types usually **typedef**-ed from standard C types. Basic types are defined in a few files. Examples are shown on this slide.

Call Types (x64)

⦿ **Direct** (same executable or shared library)

```
shell_execve:
...
0x000055e16508f807 <+103>:   callq  0x55e1650891a0 <file_isdir>
```

⦿ **Indirect**

   • Pointer (`.got.plt`)

      ○ **PLT** inter-module call

```
shell_execve:
...
0x000055e16508f89b <+251>:   callq  0x55e165078aa0 <open@plt>

open@plt:
...
0x000055e165078aa0 <+0>: jmpq   *0xe7aaa(%rip)        # 0x55e165160550 <open@got.plt>
```

Now we come from source code to implementation details. Linux API calls are ultimately implemented as call instructions on Intel and branch and link instructions on ARM platforms. In our training, we show examples in Exercises for x64 and ARM64. The compiler chooses different variants of call instructions based on whether the destination is in the same module or not. Direct calls may be emitted when a callee in the same module, a relative offset to the current instruction pointer is known, or an address is absolute.

Call Types (A64)

- ◉ Direct (same executable or shared library)

```
shell_execve:
...
0x0000aaaabb6f71e4 <+160>:    bl      0xaaaabb742530 <file_status>
```

- ◉ Indirect

  - • Pointer (.got)

    - ○ PLT inter-module call

```
shell_execve:
...
0x0000aaaabb6f7210 <+204>:    bl      0xaaaabb6d1ec0 <open@plt>

open@plt:
...
0x0000aaaabb6d1ec0 <+0>:      adrp    x16, 0xaaaabb7ff000
0x0000aaaabb6d1ec4 <+4>:      ldr     x17, [x16, #1464]
...
0x0000aaaabb6d1ecc <+12>:     br      x17
```

In ARM64, we see a similar picture where GOT (Global Offset Table) is used. We also see an example of intra-procedure registers usage (**X16** and **X17**).

API as Interface

- Provided by (exported from) some .so library

- Used by (imported by) executable or .so

- Can be functional or object-oriented

```
┌─────────────────────┐            ┌─────────────────────┐
│                     │            │                     │
│   executable/.so    │────●──────▶│        .so          │
│                     │            │                     │
└─────────────────────┘            └─────────────────────┘
```

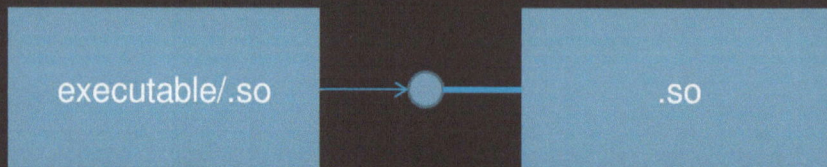

Most of the time, Linux API means a set of interfaces provided by or exported from some Shared Library. These interfaces are used by or imported by some other Shared Library or executable. Interfaces can be functional or object-oriented conceptually. Let's look at how these API interfaces are implemented.

Exploration Tools

- ⊙ <u>ldd</u>

- ⊙ <u>readelf</u>

- ⊙ <u>objdump</u>

- ⊙ **$** <u>LD_DEBUG</u>=libs bash

In addition to tracing and GDB, there are several exploration tools for shared library interfaces and dependencies that we will use in some exercises.

ldd
https://man7.org/linux/man-pages/man1/ldd.1.html

readelf
https://man7.org/linux/man-pages/man1/readelf.1.html

objdump
https://man7.org/linux/man-pages/man1/objdump.1.html

LD_DEBUG
https://man7.org/linux/man-pages/man8/ld.so.8.html

Symbol Import/Export

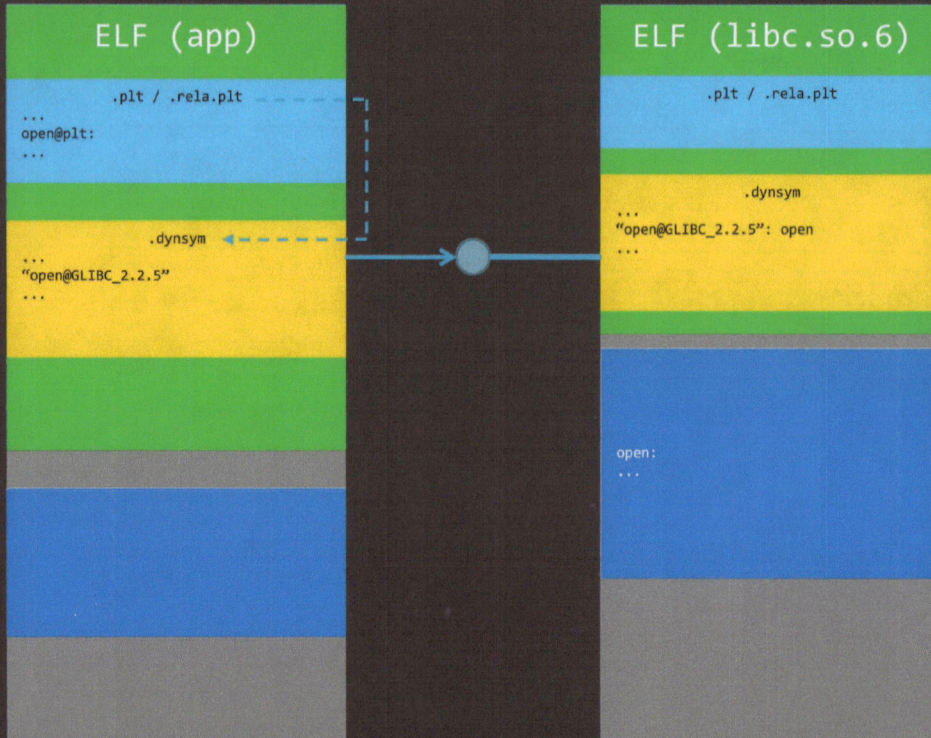

© 2023 Software Diagnostics Services

A shared library that wants to provide an interface has the so-called **.dynsym** section of names and corresponding locations in code. On the other side, if some shared library or executable file, wants to use some names from that interface, it also has the so-called **.rela.plt** section that references its own **.dynsym** section that contains these names, as illustrated on this slide where *app* wants to use *open* from *libc.so.6*.

Procedure Linkage Table (x64)

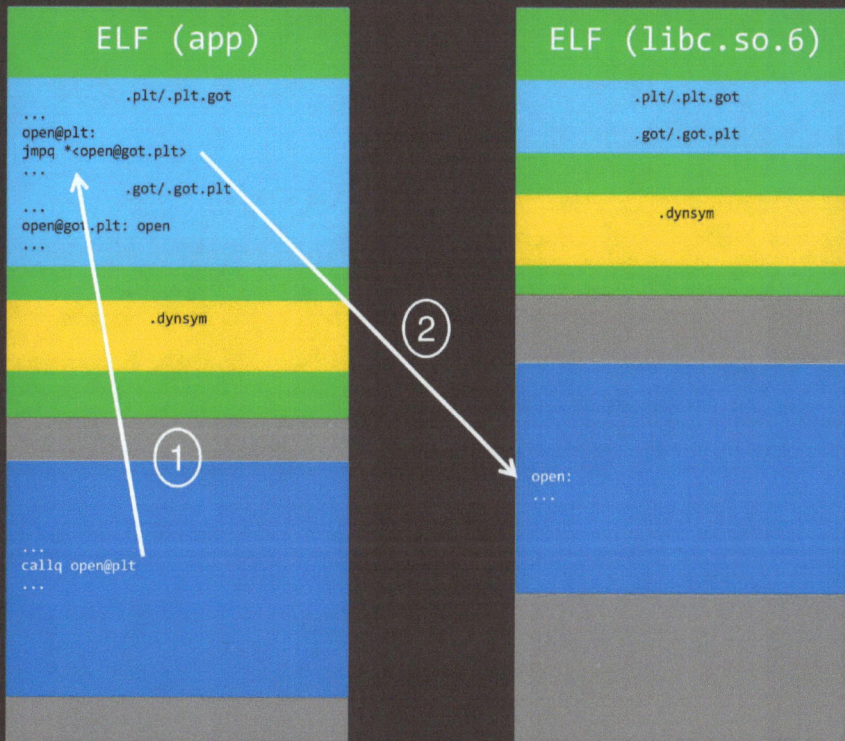

When a shared library or executable that wants to use interfaces from other shared libraries is loaded, the dynamic linker modifies the so-called **.GOT.PLT** section by populating each entry there with the real addresses from other already loaded shared libraries that export the required interface entries. For example, if the *app* wanted to use the *open* function from *libc.so.6*, the corresponding entry in **.GOT.PLT** will be filled with the real address of that function inside the *libc.so.6* code. If we want to call an interface entry, indirect addressing is used to transfer execution to the interface shared library code. First, it calls an entry in **.PLT** stub and then jumps indirectly to an address from **.GOT.PLT** that points to code in another shared library that implements the interface entry. This indirection allows calling interfaces independently from their implementation code location in memory since the relative address of **.PLT** is the same. This mechanism also allows shared libraries to be loaded at different addresses during each program run, as illustrated in the next slide. The reason why two sections **.PLT** and **.GOT.PLT** are used will be clear when we look at dynamic linking later in this course.

Procedure Linkage Table (A64)

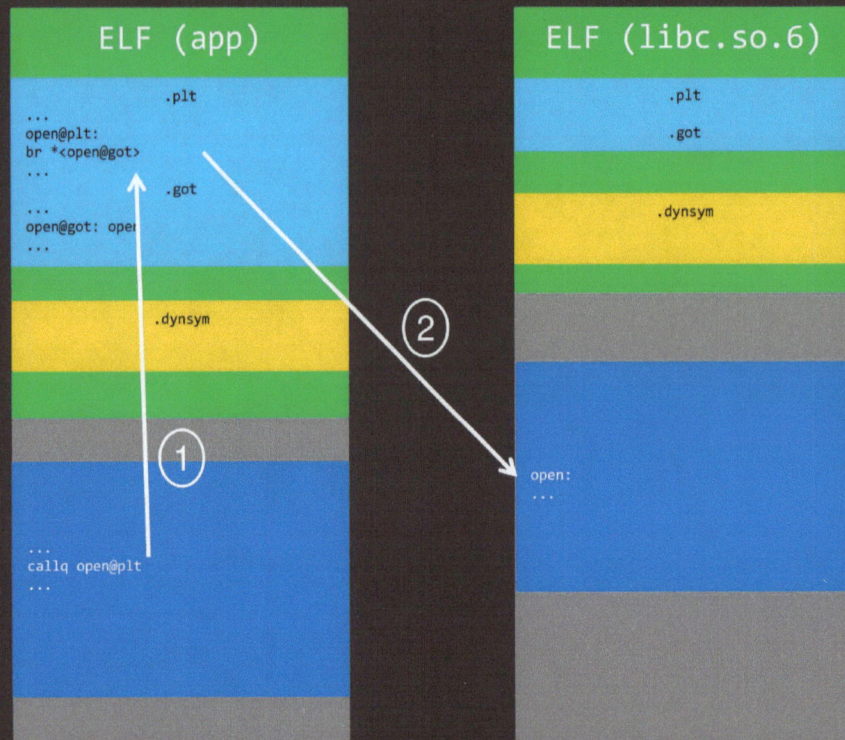

The picture for ARM64 is similar, except that section names are slightly different.

Virtual Process Address Space

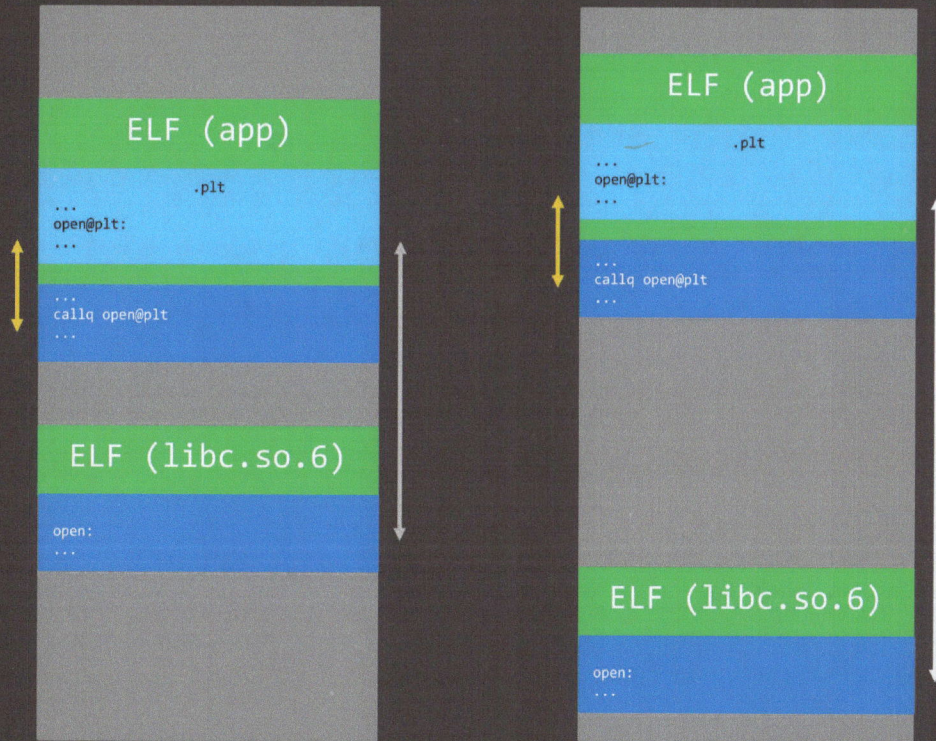

This slide shows different program executions where, each time, the same shared library is loaded at different addresses. We see that indirect addressing allows the *app* code to call the *libc* code loaded at different addresses in virtual memory.

Exercise L1

- **Goal:** Explore import/export information and calls to shared libraries

- **ADDR Patterns:** Call Path

- \LAPI-Dumps\Exercise-L1-GDB.pdf

- \LAPI-Dumps\Exercise-L1-WinDbg.pdf

Exercise L1 (GDB)

Goal: Explore import/export information and calls to shared libraries.

ADDR Patterns: Call Path.

1.   Disassemble the *execve@plt* function from the bash executable in the x64 directory:

```
~/LAPI/x64$ objdump -d bash | grep -A 4 "<execve@plt>:"
000000000002d5e0 <execve@plt>:
  2d5e0:       ff 25 0a 7d 0e 00       jmpq   *0xe7d0a(%rip)          # 1152f0 <execve@GLIBC_2.2.5>
  2d5e6:       68 5b 00 00 00          pushq  $0x5b
  2d5eb:       e9 30 fa ff ff          jmpq   2d020 <endgrent@plt-0x10>
```

2.   Dump all ELF info from the *bash* executable in the x64 directory and look for the 1152f0 entry in the *.rela.plt* section and the referenced 94 (0x5e) entry in the *.dynsym* section:

```
~/LAPI/x64$ readelf -a bash
[...]
Relocation section '.rela.plt' at offset 0x2b928 contains 218 entries:
  Offset          Info           Type           Sym. Value    Sym. Name + Addend
[...]
0000001152f0  005e00000007 R_X86_64_JUMP_SLO 0000000000000000 execve@GLIBC_2.2.5 + 0
[...]
Symbol table '.dynsym' contains 2432 entries:
   Num:    Value          Size Type    Bind   Vis      Ndx Name
[...]
    94: 0000000000000000     0 FUNC    GLOBAL DEFAULT  UND execve@GLIBC_2.2.5 (2)
[...]
```

3.   Dump all ELF info from the *libc.so.6* shared library in the x64 directory and look for the *execve@GLIBC* function in the *.dynsym* section:

```
~/LAPI/x64$ readelf -a libc.so.6
[...]
Symbol table '.dynsym' contains 2359 entries:
   Num:    Value          Size Type    Bind   Vis      Ndx Name
[...]
  1508: 00000000000c68a0    33 FUNC    WEAK   DEFAULT   13 execve@@GLIBC_2.2.5
```

4.   Load a core dump *core.9* and *bash* executable from the x64 directory:

```
~/LAPI/x64$ gdb -c core.9 -se bash
GNU gdb (Debian 8.2.1-2+b3) 8.2.1
Copyright (C) 2018 Free Software Foundation, Inc.
License GPLv3+: GNU GPL version 3 or later <http://gnu.org/licenses/gpl.html>
This is free software: you are free to change and redistribute it.
There is NO WARRANTY, to the extent permitted by law.
Type "show copying" and "show warranty" for details.
This GDB was configured as "x86_64-linux-gnu".
Type "show configuration" for configuration details.
For bug reporting instructions, please see:
<http://www.gnu.org/software/gdb/bugs/>.
Find the GDB manual and other documentation resources online at:
```

<http://www.gnu.org/software/gdb/documentation/>.

For help, type "help".
Type "apropos word" to search for commands related to "word"...
Reading symbols from bash...(no debugging symbols found)...done.

warning: core file may not match specified executable file.
[New LWP 9]
Core was generated by `-bash'.
#0 0x00007f3e9f7492d7 in __GI___waitpid (pid=-1, stat_loc=0x7ffcbd661ad0, options=10) at
../sysdeps/unix/sysv/linux/waitpid.c:30
30   ../sysdeps/unix/sysv/linux/waitpid.c: No such file or directory.

5.   Set logging to a file in case of lengthy output from some commands:

(gdb) **set logging on** L1.log
Copying output to L1.log.

6.   Disassemble the *shell_execve* function and find the call to the *open@plt* function:

(gdb) **disassemble shell_execve**
Dump of assembler code for function shell_execve:
   0x000055e16508f7a0 <+0>:   push   %r15
   0x000055e16508f7a2 <+2>:   push   %r14
   0x000055e16508f7a4 <+4>:   mov   %rdx,%r14
   0x000055e16508f7a7 <+7>:   push   %r13
   0x000055e16508f7a9 <+9>:   mov   %rsi,%r13
   0x000055e16508f7ac <+12>:   push   %r12
   0x000055e16508f7ae <+14>:   push   %rbp
   0x000055e16508f7af <+15>:   push   %rbx
   0x000055e16508f7b0 <+16>:   mov   %rdi,%rbx
   0x000055e16508f7b3 <+19>:   sub $0xa8,%rsp
   0x000055e16508f7ba <+26>:   mov   %fs:0x28,%rax
   0x000055e16508f7c3 <+35>:   mov   %rax,0x98(%rsp)
   0x000055e16508f7cb <+43>:   xor   %eax,%eax
   0x000055e16508f7cd <+45>:   callq 0x55e1650785e0 <execve@plt>
   0x000055e16508f7d2 <+50>:   callq 0x55e165078120 <__errno_location@plt>
   0x000055e16508f7d7 <+55>:   mov   %rax,%r12
   0x000055e16508f7da <+58>:   mov   (%rax),%ebp
   0x000055e16508f7dc <+60>:   mov   0xdc2a6(%rip),%eax   # 0x55e16516ba88
<terminating_signal>
   0x000055e16508f7e2 <+66>:   test   %eax,%eax
   0x000055e16508f7e4 <+68>:   jne   0x55e16508f880 <shell_execve+224>
   0x000055e16508f7ea <+74>:   cmp $0x8,%ebp
   0x000055e16508f7ed <+77>:   je   0x55e16508f894 <shell_execve+244>
   0x000055e16508f7f3 <+83>:   xor   %eax,%eax
   0x000055e16508f7f5 <+85>:   cmp $0x2,%ebp
   0x000055e16508f7f8 <+88>:   mov   %rbx,%rdi
   0x000055e16508f7fb <+91>:   sete   %al
   0x000055e16508f7fe <+94>:   add $0x7e,%eax
   0x000055e16508f801 <+97>:   mov   %eax,0xde321(%rip)   # 0x55e16516db28
<last_command_exit_value>
   0x000055e16508f807 <+103>:   callq 0x55e1650891a0 <file_isdir>
   0x000055e16508f80c <+108>:   test   %eax,%eax
   0x000055e16508f80e <+110>:   jne   0x55e16508fa58 <shell_execve+696>
   0x000055e16508f814 <+116>:   mov   %rbx,%rdi
   0x000055e16508f817 <+119>:   callq 0x55e1650d3970 <executable_file>
   0x000055e16508f81c <+124>:   test   %eax,%eax
   0x000055e16508f81e <+126>:   je   0x55e16508f841 <shell_execve+161>
   0x000055e16508f820 <+128>:   cmp $0x7,%ebp

```
   0x000055e16508f823 <+131>:    je      0x55e16508f841 <shell_execve+161>
   0x000055e16508f825 <+133>:    cmp     $0xc,%ebp
   0x000055e16508f828 <+136>:    je      0x55e16508f841 <shell_execve+161>
   0x000055e16508f82a <+138>:    xor     %esi,%esi
   0x000055e16508f82c <+140>:    mov     %rbx,%rdi
   0x000055e16508f82f <+143>:    xor     %eax,%eax
   0x000055e16508f831 <+145>:    callq   0x55e165078aa0 <open@plt>
   0x000055e16508f836 <+150>:    mov     %eax,%r13d
--Type <RET> for more, q to quit, c to continue without paging--q
Quit
```

7.   Follow the call path and indirect jump to see the real call from the *libc.so.6* library:

```
(gdb) disassemble 0x55e165078aa0
Dump of assembler code for function open@plt:
   0x000055e165078aa0 <+0>:     jmpq    *0xe7aaa(%rip)          # 0x55e165160550 <open@got.plt>
   0x000055e165078aa6 <+6>:     pushq   $0xa7
   0x000055e165078aab <+11>:    jmpq    0x55e165078020
End of assembler dump.

(gdb) x/a 0x55e165160550
0x55e165160550 <open@got.plt>:  0x7f3e9f76d010 <__libc_open64>

(gdb) info sharedlibrary
From                To                  Syms Read    Shared Object Library
0x00007f3e9f856950  0x00007f3e9f863dc8  Yes (*)      /lib/x86_64-linux-gnu/libtinfo.so.6
0x00007f3e9f844130  0x00007f3e9f844eb5  Yes          /lib/x86_64-linux-gnu/libdl.so.2
0x00007f3e9f6a5320  0x00007f3e9f7eb14b  Yes          /lib/x86_64-linux-gnu/libc.so.6
0x00007f3e9f884090  0x00007f3e9f8a1b50  Yes          /lib64/ld-linux-x86-64.so.2
0x00007f3e9f37e300  0x00007f3e9f384578  Yes          /lib/x86_64-linux-gnu/libnss_files.so.2
(*): Shared library is missing debugging information.
```

8.   We can also check the section for the address 0x55e165160550:

```
(gdb) info file
Symbols from "/home/coredump/LAPI/x64/bash".
Local core dump file:
        `/home/coredump/LAPI/x64/core.9', file type elf64-x86-64.
        0x000055e16515d000 - 0x000055e165160000 is load1
        0x000055e165160000 - 0x000055e165169000 is load2
        0x000055e165169000 - 0x000055e165173000 is load3
        0x000055e1667f5000 - 0x000055e166879000 is load4
        0x00007f3e9f388000 - 0x00007f3e9f389000 is load5
        0x00007f3e9f389000 - 0x00007f3e9f38a000 is load6
        0x00007f3e9f38a000 - 0x00007f3e9f390000 is load7
        0x00007f3e9f680000 - 0x00007f3e9f683000 is load8
        0x00007f3e9f839000 - 0x00007f3e9f83d000 is load9
        0x00007f3e9f83d000 - 0x00007f3e9f83f000 is load10
        0x00007f3e9f83f000 - 0x00007f3e9f843000 is load11
        0x00007f3e9f846000 - 0x00007f3e9f847000 is load12
        0x00007f3e9f847000 - 0x00007f3e9f848000 is load13
        0x00007f3e9f871000 - 0x00007f3e9f875000 is load14
        0x00007f3e9f875000 - 0x00007f3e9f876000 is load15
        0x00007f3e9f876000 - 0x00007f3e9f878000 is load16
        0x00007f3e9f8aa000 - 0x00007f3e9f8ab000 is load17
        0x00007f3e9f8ab000 - 0x00007f3e9f8ac000 is load18
        0x00007f3e9f8ac000 - 0x00007f3e9f8ad000 is load19
        0x00007ffcbd642000 - 0x00007ffcbd663000 is load20
        0x00007ffcbd7ea000 - 0x00007ffcbd7eb000 is load21
Local exec file:
        `/home/coredump/LAPI/x64/bash', file type elf64-x86-64.
        Entry point: 0x55e16507a630
        0x000055e16504b2a8 - 0x000055e16504b2c4 is .interp
        0x000055e16504b2c4 - 0x000055e16504b2e4 is .note.ABI-tag
        0x000055e16504b2e4 - 0x000055e16504b308 is .note.gnu.build-id
```

```
        0x000055e16504b308 - 0x000055e16504fdb0 is .gnu.hash
        0x000055e16504fdb0 - 0x000055e16505e1b0 is .dynsym
        0x000055e16505e1b0 - 0x000055e1650678d8 is .dynstr
        0x000055e1650678d8 - 0x000055e165068bd8 is .gnu.version
        0x000055e165068bd8 - 0x000055e165068ca8 is .gnu.version_r
        0x000055e165068ca8 - 0x000055e165076928 is .rela.dyn
        0x000055e165076928 - 0x000055e165077d98 is .rela.plt
        0x000055e165078000 - 0x000055e165078017 is .init
        0x000055e165078020 - 0x000055e165078dd0 is .plt
        0x000055e165078dd0 - 0x000055e165078de8 is .plt.got
        0x000055e165078df0 - 0x000055e165125781 is .text
        0x000055e165125784 - 0x000055e16512578d is .fini
        0x000055e165126000 - 0x000055e16513f930 is .rodata
        0x000055e16513f930 - 0x000055e165143df4 is .eh_frame_hdr
        0x000055e165143df8 - 0x000055e16515b730 is .eh_frame
        0x000055e16515d3f0 - 0x000055e16515d3f8 is .init_array
        0x000055e16515d3f8 - 0x000055e16515d400 is .fini_array
        0x000055e16515d400 - 0x000055e16515fcf0 is .data.rel.ro
--Type <RET> for more, q to quit, c to continue without paging--
        0x000055e16515fcf0 - 0x000055e16515fef0 is .dynamic
        0x000055e16515fef0 - 0x000055e16515fff0 is .got
        0x000055e165160000 - 0x000055e1651606e8 is .got.plt
        0x000055e165160700 - 0x000055e165168d04 is .data
        0x000055e165168d20 - 0x000055e165172998 is .bss
        0x00007f3e9f848238 - 0x00007f3e9f84825c is .note.gnu.build-id in /lib/x86_64-linux-gnu/libtinfo.so.6
        0x00007f3e9f848260 - 0x00007f3e9f848a78 is .gnu.hash in /lib/x86_64-linux-gnu/libtinfo.so.6
        0x00007f3e9f848a78 - 0x00007f3e9f84a7a0 is .dynsym in /lib/x86_64-linux-gnu/libtinfo.so.6
        0x00007f3e9f84a7a0 - 0x00007f3e9f84b71a is .dynstr in /lib/x86_64-linux-gnu/libtinfo.so.6
        0x00007f3e9f84b71a - 0x00007f3e9f84b988 is .gnu.version in /lib/x86_64-linux-gnu/libtinfo.so.6
        0x00007f3e9f84b988 - 0x00007f3e9f84bd34 is .gnu.version_d in /lib/x86_64-linux-gnu/libtinfo.so.6
        0x00007f3e9f84bd38 - 0x00007f3e9f84bda8 is .gnu.version_r in /lib/x86_64-linux-gnu/libtinfo.so.6
        0x00007f3e9f84bda8 - 0x00007f3e9f854d48 is .rela.dyn in /lib/x86_64-linux-gnu/libtinfo.so.6
        0x00007f3e9f854d48 - 0x00007f3e9f855ae0 is .rela.plt in /lib/x86_64-linux-gnu/libtinfo.so.6
        0x00007f3e9f856000 - 0x00007f3e9f856017 is .init in /lib/x86_64-linux-gnu/libtinfo.so.6
        0x00007f3e9f856020 - 0x00007f3e9f856940 is .plt in /lib/x86_64-linux-gnu/libtinfo.so.6
        0x00007f3e9f856940 - 0x00007f3e9f856950 is .plt.got in /lib/x86_64-linux-gnu/libtinfo.so.6
        0x00007f3e9f856950 - 0x00007f3e9f863dc8 is .text in /lib/x86_64-linux-gnu/libtinfo.so.6
        0x00007f3e9f863dc8 - 0x00007f3e9f863dd1 is .fini in /lib/x86_64-linux-gnu/libtinfo.so.6
        0x00007f3e9f864000 - 0x00007f3e9f86d9a5 is .rodata in /lib/x86_64-linux-gnu/libtinfo.so.6
        0x00007f3e9f86d9a8 - 0x00007f3e9f86e13c is .eh_frame_hdr in /lib/x86_64-linux-gnu/libtinfo.so.6
        0x00007f3e9f86e140 - 0x00007f3e9f8706b0 is .eh_frame in /lib/x86_64-linux-gnu/libtinfo.so.6
        0x00007f3e9f8718d0 - 0x00007f3e9f8718d8 is .init_array in /lib/x86_64-linux-gnu/libtinfo.so.6
        0x00007f3e9f8718d8 - 0x00007f3e9f8718e0 is .fini_array in /lib/x86_64-linux-gnu/libtinfo.so.6
        0x00007f3e9f8718e0 - 0x00007f3e9f874848 is .data.rel.ro in /lib/x86_64-linux-gnu/libtinfo.so.6
        0x00007f3e9f874848 - 0x00007f3e9f874a58 is .dynamic in /lib/x86_64-linux-gnu/libtinfo.so.6
        0x00007f3e9f874a58 - 0x00007f3e9f874ff8 is .got in /lib/x86_64-linux-gnu/libtinfo.so.6
        0x00007f3e9f875000 - 0x00007f3e9f8754f4 is .data in /lib/x86_64-linux-gnu/libtinfo.so.6
        0x00007f3e9f875500 - 0x00007f3e9f875980 is .bss in /lib/x86_64-linux-gnu/libtinfo.so.6
        0x00007f3e9f843238 - 0x00007f3e9f84325c is .note.gnu.build-id in /lib/x86_64-linux-gnu/libdl.so.2
        0x00007f3e9f84325c - 0x00007f3e9f84327c is .note.ABI-tag in /lib/x86_64-linux-gnu/libdl.so.2
        0x00007f3e9f843280 - 0x00007f3e9f843348 is .gnu.hash in /lib/x86_64-linux-gnu/libdl.so.2
        0x00007f3e9f843348 - 0x00007f3e9f843750 is .dynsym in /lib/x86_64-linux-gnu/libdl.so.2
        0x00007f3e9f843750 - 0x00007f3e9f843989 is .dynstr in /lib/x86_64-linux-gnu/libdl.so.2
        0x00007f3e9f84398a - 0x00007f3e9f8439e0 is .gnu.version in /lib/x86_64-linux-gnu/libdl.so.2
        0x00007f3e9f8439e0 - 0x00007f3e9f843a84 is .gnu.version_d in /lib/x86_64-linux-gnu/libdl.so.2
        0x00007f3e9f843a88 - 0x00007f3e9f843ae8 is .gnu.version_r in /lib/x86_64-linux-gnu/libdl.so.2
        0x00007f3e9f843ae8 - 0x00007f3e9f843c80 is .rela.dyn in /lib/x86_64-linux-gnu/libdl.so.2
        0x00007f3e9f843c80 - 0x00007f3e9f843db8 is .rela.plt in /lib/x86_64-linux-gnu/libdl.so.2
        0x00007f3e9f844000 - 0x00007f3e9f844017 is .init in /lib/x86_64-linux-gnu/libdl.so.2
        0x00007f3e9f844020 - 0x00007f3e9f844100 is .plt in /lib/x86_64-linux-gnu/libdl.so.2
        0x00007f3e9f844100 - 0x00007f3e9f844128 is .plt.got in /lib/x86_64-linux-gnu/libdl.so.2
        0x00007f3e9f844130 - 0x00007f3e9f844eb5 is .text in /lib/x86_64-linux-gnu/libdl.so.2
        0x00007f3e9f844eb8 - 0x00007f3e9f844ec1 is .fini in /lib/x86_64-linux-gnu/libdl.so.2
        0x00007f3e9f845000 - 0x00007f3e9f8450a3 is .rodata in /lib/x86_64-linux-gnu/libdl.so.2
--Type <RET> for more, q to quit, c to continue without paging—q
Quit
```

9.   Load a core dump *core.19649* and *bash* executable from the A64 directory:

```
~/LAPI/x64$ cd ../A64
```

```
~/LAPI/A64$ gdb-multiarch -c core.19649 -se bash
GNU gdb (Debian 8.2.1-2+b3) 8.2.1
```

For help, type "help".
Type "apropos word" to search for commands related to "word"...
Reading symbols from bash...(no debugging symbols found)...done.

warning: core file may not match specified executable file.
[New LWP 19649]

warning: Could not load shared library symbols for 3 libraries, e.g.
/lib/aarch64-linux-gnu/libtinfo.so.6.
Use the "info sharedlibrary" command to see the complete listing.
Do you need "set solib-search-path" or "set sysroot"?
Core was generated by `-bash'.
#0 0x0000ffffbafa6734 in ?? ()

(gdb) **set solib-search-path** .
Reading symbols from /home/coredump/LAPI/A64/libtinfo.so.6...(no debugging symbols found)...done.
Reading symbols from /home/coredump/LAPI/A64/libc.so.6...(no debugging symbols found)...done.
Reading symbols from /home/coredump/LAPI/A64/ld-linux-aarch64.so.1...(no debugging symbols found)...done.

10.	Set logging to a file in case of lengthy output from some commands:

(gdb) **set logging on** L1.log
Copying output to L1.log.

11.	Disassemble the *shell_execve* function and find the call to the *open@plt* function:

(gdb) **disassemble shell_execve**
Dump of assembler code for function shell_execve:
```
   0x0000aaaabb6f7144 <+0>:     stp     x29, x30, [sp, #-224]!
   0x0000aaaabb6f7148 <+4>:     adrp    x3, 0xaaaabb800000
   0x0000aaaabb6f714c <+8>:     mov     x29, sp
   0x0000aaaabb6f7150 <+12>:    ldr     x3, [x3, #1192]
   0x0000aaaabb6f7154 <+16>:    stp     x19, x20, [sp, #16]
   0x0000aaaabb6f7158 <+20>:    mov     x19, x0
   0x0000aaaabb6f715c <+24>:    stp     x21, x22, [sp, #32]
   0x0000aaaabb6f7160 <+28>:    mov     x22, x1
   0x0000aaaabb6f7164 <+32>:    stp     x23, x24, [sp, #48]
   0x0000aaaabb6f7168 <+36>:    mov     x23, x2
   0x0000aaaabb6f716c <+40>:    stp     x25, x26, [sp, #64]
   0x0000aaaabb6f7170 <+44>:    ldr     x4, [x3]
   0x0000aaaabb6f7174 <+48>:    str     x4, [sp, #216]
   0x0000aaaabb6f7178 <+52>:    mov     x4, #0x0                      // #0
   0x0000aaaabb6f717c <+56>:    bl      0xaaaabb6d23e0 <execve@plt>
   0x0000aaaabb6f7180 <+60>:    bl      0xaaaabb6d2740 <__errno_location@plt>
   0x0000aaaabb6f7184 <+64>:    mov     x20, x0
   0x0000aaaabb6f7188 <+68>:    adrp    x3, 0xaaaabb800000
   0x0000aaaabb6f718c <+72>:    ldr     x0, [x3, #2616]
   0x0000aaaabb6f7190 <+76>:    ldr     w1, [x0]
```

81

```
   0x0000aaaabb6f7194 <+80>:     cbnz    w1, 0xaaaabb6f72bc <shell_execve+376>
   0x0000aaaabb6f7198 <+84>:     ldr     w21, [x20]
   0x0000aaaabb6f719c <+88>:     cmp     w21, #0x8
   0x0000aaaabb6f71a0 <+92>:     b.eq    0xaaaabb6f73c4 <shell_execve+640>   // b.none
   0x0000aaaabb6f71a4 <+96>:     adrp    x23, 0xaaaabb800000
   0x0000aaaabb6f71a8 <+100>:    cmp     w21, #0x2
   0x0000aaaabb6f71ac <+104>:    cset    w2, eq  // eq = none
   0x0000aaaabb6f71b0 <+108>:    add     x24, sp, #0x58
   0x0000aaaabb6f71b4 <+112>:    ldr     x3, [x23, #3744]
   0x0000aaaabb6f71b8 <+116>:    add     w2, w2, #0x7e
   0x0000aaaabb6f71bc <+120>:    mov     x1, x24
   0x0000aaaabb6f71c0 <+124>:    mov     x0, x19
   0x0000aaaabb6f71c4 <+128>:    str     w2, [x3]
   0x0000aaaabb6f71c8 <+132>:    bl      0xaaaabb6d2180 <stat@plt>
   0x0000aaaabb6f71cc <+136>:    cbnz    w0, 0xaaaabb6f71e0 <shell_execve+156>
   0x0000aaaabb6f71d0 <+140>:    ldr     w0, [sp, #104]
   0x0000aaaabb6f71d4 <+144>:    and     w0, w0, #0xf000
   0x0000aaaabb6f71d8 <+148>:    cmp     w0, #0x4, lsl #12
   0x0000aaaabb6f71dc <+152>:    b.eq    0xaaaabb6f7288 <shell_execve+324>   // b.none
   0x0000aaaabb6f71e0 <+156>:    mov     x0, x19
   0x0000aaaabb6f71e4 <+160>:    bl      0xaaaabb742530 <file_status>
   0x0000aaaabb6f71e8 <+164>:    tbnz    w0, #4, 0xaaaabb6f727c <shell_execve+312>
   0x0000aaaabb6f71ec <+168>:    mov     w1, #0x12                            // #18
   0x0000aaaabb6f71f0 <+172>:    and     w22, w0, w1
--Type <RET> for more, q to quit, c to continue without paging--
   0x0000aaaabb6f71f4 <+176>:    cmp     w22, #0x2
   0x0000aaaabb6f71f8 <+180>:    b.ne    0xaaaabb6f721c <shell_execve+216>   // b.any
   0x0000aaaabb6f71fc <+184>:    cmp     w21, #0x7
   0x0000aaaabb6f7200 <+188>:    ccmp    w21, #0xc, #0x4, ne  // ne = any
   0x0000aaaabb6f7204 <+192>:    b.eq    0xaaaabb6f721c <shell_execve+216>   // b.none
   0x0000aaaabb6f7208 <+196>:    mov     x0, x19
   0x0000aaaabb6f720c <+200>:    mov     w1, #0x0                             // #0
   0x0000aaaabb6f7210 <+204>:    bl      0xaaaabb6d1ec0 <open@plt>
   0x0000aaaabb6f7214 <+208>:    mov     w25, w0
   0x0000aaaabb6f7218 <+212>:    tbz     w0, #31, 0xaaaabb6f72d0 <shell_execve+396>
   0x0000aaaabb6f721c <+216>:    str     w21, [x20]
   0x0000aaaabb6f7220 <+220>:    mov     w0, w21
   0x0000aaaabb6f7224 <+224>:    bl      0xaaaabb6d2110 <strerror@plt>
   0x0000aaaabb6f7228 <+228>:    mov     x2, x0
   0x0000aaaabb6f722c <+232>:    mov     x1, x19
   0x0000aaaabb6f7230 <+236>:    adrp    x0, 0xaaaabb7c4000
   0x0000aaaabb6f7234 <+240>:    add     x0, x0, #0xbd8
   0x0000aaaabb6f7238 <+244>:    bl      0xaaaabb706674 <report_error>
   0x0000aaaabb6f723c <+248>:    ldr     x23, [x23, #3744]
   0x0000aaaabb6f7240 <+252>:    ldr     w25, [x23]
   0x0000aaaabb6f7244 <+256>:    adrp    x0, 0xaaaabb800000
   0x0000aaaabb6f7248 <+260>:    ldr     x0, [x0, #1192]
   0x0000aaaabb6f724c <+264>:    ldr     x2, [sp, #216]
   0x0000aaaabb6f7250 <+268>:    ldr     x1, [x0]
   0x0000aaaabb6f7254 <+272>:    subs    x2, x2, x1
   0x0000aaaabb6f7258 <+276>:    mov     x1, #0x0                             // #0
   0x0000aaaabb6f725c <+280>:    b.ne    0xaaaabb6f77ec <shell_execve+1704>   // b.any
   0x0000aaaabb6f7260 <+284>:    mov     w0, w25
   0x0000aaaabb6f7264 <+288>:    ldp     x19, x20, [sp, #16]
   0x0000aaaabb6f7268 <+292>:    ldp     x21, x22, [sp, #32]
   0x0000aaaabb6f726c <+296>:    ldp     x23, x24, [sp, #48]
   0x0000aaaabb6f7270 <+300>:    ldp     x25, x26, [sp, #64]
   0x0000aaaabb6f7274 <+304>:    ldp     x29, x30, [sp], #224
   0x0000aaaabb6f7278 <+308>:    ret
   0x0000aaaabb6f727c <+312>:    mov     w1, #0x15                            // #21
```

```
    0x0000aaaabb6f7280 <+316>:    str     w1, [x20]
    0x0000aaaabb6f7284 <+320>:    b       0xaaaabb6f71ec <shell_execve+168>
    0x0000aaaabb6f7288 <+324>:    adrp    x1, 0xaaaabb7c4000
    0x0000aaaabb6f728c <+328>:    add     x1, x1, #0xbd8
    0x0000aaaabb6f7290 <+332>:    mov     w2, #0x5                        // #5
    0x0000aaaabb6f7294 <+336>:    mov     x0, #0x0                        // #0
    0x0000aaaabb6f7298 <+340>:    bl      0xaaaabb6d2630 <dcgettext@plt>
    0x0000aaaabb6f729c <+344>:    mov     x20, x0
    0x0000aaaabb6f72a0 <+348>:    mov     w0, #0x15                       // #21
    0x0000aaaabb6f72a4 <+352>:    bl      0xaaaabb6d2110 <strerror@plt>
--Type <RET> for more, q to quit, c to continue without paging--q
Quit
```

12.   Follow the call path to see the real branch and link to the *libc.so.6* library:

```
(gdb) disassemble 0xaaaabb6d1ec0
Dump of assembler code for function open@plt:
    0x0000aaaabb6d1ec0 <+0>:     adrp    x16, 0xaaaabb7ff000
    0x0000aaaabb6d1ec4 <+4>:     ldr     x17, [x16, #1464]
    0x0000aaaabb6d1ec8 <+8>:     add     x16, x16, #0x5b8
    0x0000aaaabb6d1ecc <+12>:    br      x17
End of assembler dump.
```

```
(gdb) x/a 0xaaaabb7ff000+1464
0xaaaabb7ff5b8 <open@got.plt>:  0xffffbafc7810 <open64>
```

```
(gdb) info sharedlibrary
From                To                  Syms Read     Shared Object Library
0x0000ffffbb0ad860  0x0000ffffbb0bcb88  Yes (*)       /home/coredump/LAPI/A64/libtinfo.so.6
0x0000ffffbaf17040  0x0000ffffbb023f20  Yes (*)       /home/coredump/LAPI/A64/libc.so.6
0x0000ffffbb0eec40  0x0000ffffbb10d064  Yes (*)       /home/coredump/LAPI/A64/ld-linux-
aarch64.so.1
(*): Shared library is missing debugging information.
```

13.   We can also check the section for the address 0xaaaabb7ff000+1464 (0xaaaabb7ff5b8):

```
(gdb) info file
Symbols from "/home/coredump/LAPI/A64/bash".
Local core dump file:
        `/home/coredump/LAPI/A64/core.19649', file type elf64-littleaarch64.
        0x0000aaaabb6a0000 - 0x0000aaaabb7ed000 is load1
        0x0000aaaabb7fc000 - 0x0000aaaabb801000 is load2
        0x0000aaaabb801000 - 0x0000aaaabb80a000 is load3
        0x0000aaaabb80a000 - 0x0000aaaabb815000 is load4
        0x0000aaaae4664000 - 0x0000aaaae4815000 is load5
        0x0000ffffbaef0000 - 0x0000ffffbb079000 is load6
        0x0000ffffbb088000 - 0x0000ffffbb08c000 is load7
        0x0000ffffbb08c000 - 0x0000ffffbb08e000 is load8
        0x0000ffffbb08e000 - 0x0000ffffbb09a000 is load9
        0x0000ffffbb0a0000 - 0x0000ffffbb0cc000 is load10
        0x0000ffffbb0db000 - 0x0000ffffbb0df000 is load11
        0x0000ffffbb0df000 - 0x0000ffffbb0e0000 is load12
        0x0000ffffbb0ed000 - 0x0000ffffbb118000 is load13
        0x0000ffffbb119000 - 0x0000ffffbb11b000 is load14
        0x0000ffffbb122000 - 0x0000ffffbb124000 is load15
        0x0000ffffbb126000 - 0x0000ffffbb127000 is load16
        0x0000ffffbb127000 - 0x0000ffffbb129000 is load17
        0x0000ffffbb129000 - 0x0000ffffbb12b000 is load18
        0x0000ffffc5fb8000 - 0x0000ffffc5fd9000 is load19
```

```
Local exec file:
        `/home/coredump/LAPI/A64/bash', file type elf64-littleaarch64.
        Entry point: 0xaaaabb6d4440
        0x0000aaaabb6a0238 - 0x0000aaaabb6a0253 is .interp
        0x0000aaaabb6a0254 - 0x0000aaaabb6a0278 is .note.gnu.build-id
        0x0000aaaabb6a0278 - 0x0000aaaabb6a0298 is .note.ABI-tag
        0x0000aaaabb6a0298 - 0x0000aaaabb6a4e54 is .gnu.hash
        0x0000aaaabb6a4e58 - 0x0000aaaabb6b3a50 is .dynsym
        0x0000aaaabb6b3a50 - 0x0000aaaabb6bd6f7 is .dynstr
        0x0000aaaabb6bd6f8 - 0x0000aaaabb6beaa2 is .gnu.version
        0x0000aaaabb6beaa8 - 0x0000aaaabb6beb38 is .gnu.version_r
        0x0000aaaabb6beb38 - 0x0000aaaabb6d0538 is .rela.dyn
        0x0000aaaabb6d0538 - 0x0000aaaabb6d1a50 is .rela.plt
        0x0000aaaabb6d1a50 - 0x0000aaaabb6d1a68 is .init
        0x0000aaaabb6d1a70 - 0x0000aaaabb6d28a0 is .plt
        0x0000aaaabb6d28c0 - 0x0000aaaabb7af8b0 is .text
        0x0000aaaabb7af8b0 - 0x0000aaaabb7af8c4 is .fini
        0x0000aaaabb7af8c8 - 0x0000aaaabb7c8f24 is .rodata
        0x0000aaaabb7c8f24 - 0x0000aaaabb7cd4f0 is .eh_frame_hdr
        0x0000aaaabb7cd4f0 - 0x0000aaaabb7ec480 is .eh_frame
        0x0000aaaabb7fc818 - 0x0000aaaabb7fc820 is .init_array
        0x0000aaaabb7fc820 - 0x0000aaaabb7fc828 is .fini_array
        0x0000aaaabb7fc828 - 0x0000aaaabb7ff178 is .data.rel.ro
        0x0000aaaabb7ff178 - 0x0000aaaabb7ff388 is .dynamic
        0x0000aaaabb7ff388 - 0x0000aaaabb801000 is .got
        0x0000aaaabb801000 - 0x0000aaaabb809280 is .data
--Type <RET> for more, q to quit, c to continue without paging--q
Quit
```

Note: We see that for ARM64 implementation it is .GOT (Global Offset Table) section.

Exercise L1 (WinDbg)

Goal: Explore import/export information and calls to shared libraries.

ADDR Patterns: Call Path.

1.   Launch WinDbg and load a core dump *core.19649* from the A64 directory:

```
Microsoft (R) Windows Debugger Version 10.0.25324.1001 AMD64
Copyright (c) Microsoft Corporation. All rights reserved.


Loading Dump File [C:\LAPI\A64\core.19649]
64-bit machine not using 64-bit API

************* Path validation summary **************
Response                      Time (ms)      Location
Deferred                                     srv*
Symbol search path is: srv*
Executable search path is:
Generic Unix Version 0 UP Free ARM 64-bit (AArch64)
System Uptime: not available
Process Uptime: not available
.................
*** WARNING: Unable to verify timestamp for libc.so.6
*** WARNING: Unable to verify timestamp for bash
libc_so+0xb6734:
0000ffff`bafa6734 d4000001 svc          #0
```

2.   Set the symbol path and logging to a file in case of lengthy output from some commands:

```
0:000> .sympath+ C:\LAPI\A64
Symbol search path is: srv*;C:\LAPI\A64
Expanded Symbol search path is: cache*;SRV*https://msdl.microsoft.com/download/symbols;c:\
lapi\a64


************* Path validation summary **************
Response                      Time (ms)      Location
Deferred                                     srv*
OK                                           C:\LAPI\A64
*** WARNING: Unable to verify timestamp for libc.so.6
*** WARNING: Unable to verify timestamp for bash

0:000> .reload
.....*** WARNING: Unable to verify timestamp for libc.so.6
..............
*** WARNING: Unable to verify timestamp for bash

************* Symbol Loading Error Summary **************
Module name          Error
bash                 The system cannot find the file specified
libc.so              The system cannot find the file specified

You can troubleshoot most symbol related issues by turning on symbol loading diagnostics (!sym
noisy) and repeating the command that caused symbols to be loaded.
You should also verify that your symbol search path (.sympath) is correct.
```

85

```
0:000> .logopen C:\LAPI\A64\L1-WinDbg.log
Opened log file 'C:\LAPI\A64\L1-WinDbg.log'
```

3.   Disassemble the *shell_execve* function and find the call to the address 0xaaaabb6d1ec0 identified in GDB exercises as *open@plt* function:

```
0:000> uf shell_execve
bash!shell_execve:
0000aaaa`bb6f7144 a9b27bfd stp        fp,lr,[sp,#-0xE0]!
0000aaaa`bb6f7148 b0000843 adrp       x3,bash!o_options+0x2118 (0000aaaa`bb800000)
0000aaaa`bb6f714c 910003fd mov        fp,sp
0000aaaa`bb6f7150 f9425463 ldr        x3,[x3,#0x4A8]
0000aaaa`bb6f7154 a90153f3 stp        x19,x20,[sp,#0x10]
0000aaaa`bb6f7158 aa0003f3 mov        x19,x0
0000aaaa`bb6f715c a9025bf5 stp        x21,x22,[sp,#0x20]
0000aaaa`bb6f7160 aa0103f6 mov        x22,x1
0000aaaa`bb6f7164 a90363f7 stp        x23,x24,[sp,#0x30]
0000aaaa`bb6f7168 aa0203f7 mov        x23,x2
0000aaaa`bb6f716c a9046bf9 stp        x25,x26,[sp,#0x40]
0000aaaa`bb6f7170 f9400064 ldr        x4,[x3]
0000aaaa`bb6f7174 f9006fe4 str        x4,[sp,#0xD8]
0000aaaa`bb6f7178 d2800004 mov        x4,#0
0000aaaa`bb6f717c 97ff6c99 bl         bash+0x323e0 (0000aaaa`bb6d23e0)
0000aaaa`bb6f7180 97ff6d70 bl         bash+0x32740 (0000aaaa`bb6d2740)
0000aaaa`bb6f7184 aa0003f4 mov        x20,x0
0000aaaa`bb6f7188 b0000843 adrp       x3,bash!o_options+0x2118 (0000aaaa`bb800000)
0000aaaa`bb6f718c f9451c60 ldr        x0,[x3,#0xA38]
0000aaaa`bb6f7190 b9400001 ldr        w1,[x0]
0000aaaa`bb6f7194 35000941 cbnz       w1,bash!shell_execve+0x178 (0000aaaa`bb6f72bc)  Branch

bash!shell_execve+0x54:
0000aaaa`bb6f7198 b9400295 ldr        w21,[x20]
0000aaaa`bb6f719c 710022bf cmp        w21,#8
0000aaaa`bb6f71a0 54001120 beq        bash!shell_execve+0x280 (0000aaaa`bb6f73c4)  Branch

bash!shell_execve+0x60:
0000aaaa`bb6f71a4 b0000857 adrp       x23,bash!o_options+0x2118 (0000aaaa`bb800000)
0000aaaa`bb6f71a8 71000abf cmp        w21,#2
0000aaaa`bb6f71ac 1a9f17e2 cseteq     w2
0000aaaa`bb6f71b0 910163f8 add        x24,sp,#0x58
0000aaaa`bb6f71b4 f94752e3 ldr        x3,[x23,#0xEA0]
0000aaaa`bb6f71b8 1101f842 add        w2,w2,#0x7E
0000aaaa`bb6f71bc aa1803e1 mov        x1,x24
0000aaaa`bb6f71c0 aa1303e0 mov        x0,x19
0000aaaa`bb6f71c4 b9000062 str        w2,[x3]
0000aaaa`bb6f71c8 97ff6bee bl         bash+0x32180 (0000aaaa`bb6d2180)
0000aaaa`bb6f71cc 350000a0 cbnz       w0,bash!shell_execve+0x9c (0000aaaa`bb6f71e0)  Branch

bash!shell_execve+0x8c:
0000aaaa`bb6f71d0 b9406be0 ldr        w0,[sp,#0x68]
0000aaaa`bb6f71d4 12140c00 and        w0,w0,#0xF000
0000aaaa`bb6f71d8 7140101f cmp        w0,#4,lsl #0xC
0000aaaa`bb6f71dc 54000560 beq        bash!shell_execve+0x144 (0000aaaa`bb6f7288)  Branch

bash!shell_execve+0x9c:
0000aaaa`bb6f71e0 aa1303e0 mov        x0,x19
0000aaaa`bb6f71e4 94012cd3 bl         bash!file_status (0000aaaa`bb742530)
0000aaaa`bb6f71e8 372004a0 tbnz       x0,#4,bash!shell_execve+0x138 (0000aaaa`bb6f727c)  Branch

bash!shell_execve+0xa8:
0000aaaa`bb6f71ec 52800241 mov        w1,#0x12
0000aaaa`bb6f71f0 0a010016 and        w22,w0,w1
0000aaaa`bb6f71f4 71000adf cmp        w22,#2
0000aaaa`bb6f71f8 54000121 bne        bash!shell_execve+0xd8 (0000aaaa`bb6f721c)  Branch

bash!shell_execve+0xb8:
0000aaaa`bb6f71fc 71001ebf cmp        w21,#7
0000aaaa`bb6f7200 7a4c1aa4 ccmpne     w21,#0xC,#4
0000aaaa`bb6f7204 540000c0 beq        bash!shell_execve+0xd8 (0000aaaa`bb6f721c)  Branch

bash!shell_execve+0xc4:
```

```
0000aaaa`bb6f7208 aa1303e0 mov        x0,x19
0000aaaa`bb6f720c 52800001 mov        w1,#0
0000aaaa`bb6f7210 97ff6b2c bl         bash+0x31ec0 (0000aaaa`bb6d1ec0)
0000aaaa`bb6f7214 2a0003f9 mov        w25,w0
0000aaaa`bb6f7218 36f805c0 tbz        x0,#0x1F,bash!shell_execve+0x18c (0000aaaa`bb6f72d0)   Branch

bash!shell_execve+0xd8:
0000aaaa`bb6f721c b9000295 str        w21,[x20]
0000aaaa`bb6f7220 2a1503e0 mov        w0,w21
0000aaaa`bb6f7224 97ff6bbb bl         bash+0x32110 (0000aaaa`bb6d2110)
0000aaaa`bb6f7228 aa0003e2 mov        x2,x0
0000aaaa`bb6f722c aa1303e1 mov        x1,x19
0000aaaa`bb6f7230 b0000660 adrp       x0,bash!build_version+0x1aa4 (0000aaaa`bb7c4000)
0000aaaa`bb6f7234 912f6000 add        x0,x0,#0xBD8
0000aaaa`bb6f7238 94003d0f bl         bash!report_error (0000aaaa`bb706674)

bash!shell_execve+0xf8:
0000aaaa`bb6f723c f94752f7 ldr        x23,[x23,#0xEA0]
0000aaaa`bb6f7240 b94002f9 ldr        w25,[x23]

bash!shell_execve+0x100:
0000aaaa`bb6f7244 b0000840 adrp       x0,bash!o_options+0x2118 (0000aaaa`bb800000)
0000aaaa`bb6f7248 f9425400 ldr        x0,[x0,#0x4A8]
0000aaaa`bb6f724c f9406fe2 ldr        x2,[sp,#0xD8]
0000aaaa`bb6f7250 f9400001 ldr        x1,[x0]
0000aaaa`bb6f7254 eb010042 subs       x2,x2,x1
0000aaaa`bb6f7258 d2800001 mov        x1,#0
0000aaaa`bb6f725c 54002c81 bne        bash!shell_execve+0x6a8 (0000aaaa`bb6f77ec)   Branch

bash!shell_execve+0x11c:
0000aaaa`bb6f7260 2a1903e0 mov        w0,w25
0000aaaa`bb6f7264 a94153f3 ldp        x19,x20,[sp,#0x10]
0000aaaa`bb6f7268 a9425bf5 ldp        x21,x22,[sp,#0x20]
0000aaaa`bb6f726c a94363f7 ldp        x23,x24,[sp,#0x30]
0000aaaa`bb6f7270 a9446bf9 ldp        x25,x26,[sp,#0x40]
0000aaaa`bb6f7274 a8ce7bfd ldp        fp,lr,[sp],#0xE0
0000aaaa`bb6f7278 d65f03c0 ret

bash!shell_execve+0x138:
0000aaaa`bb6f727c 528002a1 mov        w1,#0x15
0000aaaa`bb6f7280 b9000281 str        w1,[x20]
0000aaaa`bb6f7284 17ffffda b          bash!shell_execve+0xa8 (0000aaaa`bb6f71ec)   Branch

bash!shell_execve+0x144:
0000aaaa`bb6f7288 b0000661 adrp       x1,bash!build_version+0x1aa4 (0000aaaa`bb7c4000)
0000aaaa`bb6f728c 912f6021 add        x1,x1,#0xBD8
0000aaaa`bb6f7290 528000a2 mov        w2,#5
0000aaaa`bb6f7294 d2800000 mov        x0,#0
0000aaaa`bb6f7298 97ff6ce6 bl         bash+0x32630 (0000aaaa`bb6d2630)
0000aaaa`bb6f729c aa0003f4 mov        x20,x0
0000aaaa`bb6f72a0 528002a0 mov        w0,#0x15
0000aaaa`bb6f72a4 97ff6b9b bl         bash+0x32110 (0000aaaa`bb6d2110)
0000aaaa`bb6f72a8 aa1303e1 mov        x1,x19
0000aaaa`bb6f72ac aa0003e2 mov        x2,x0
0000aaaa`bb6f72b0 aa1403e0 mov        x0,x20
0000aaaa`bb6f72b4 940039df bl         bash!internal_error (0000aaaa`bb705a30)
0000aaaa`bb6f72b8 17fffffe1 b         bash!shell_execve+0xf8 (0000aaaa`bb6f723c)   Branch

bash!shell_execve+0x178:
0000aaaa`bb6f72bc d0000881 adrp       x1,bash!rl_color_indicator+0x128 (0000aaaa`bb809000)
0000aaaa`bb6f72c0 b9400000 ldr        w0,[x0]
0000aaaa`bb6f72c4 b9481021 ldr        w1,[x1,#0x810]
0000aaaa`bb6f72c8 35fff681 cbnz       w1,bash!shell_execve+0x54 (0000aaaa`bb6f7198)   Branch

bash!shell_execve+0x188:
0000aaaa`bb6f72cc 9400d4d1 bl         bash!sigint_sighandler+0x150 (0000aaaa`bb72c610)

bash!shell_execve+0x18c:
0000aaaa`bb6f72d0 aa1803e1 mov        x1,x24
0000aaaa`bb6f72d4 d2801002 mov        x2,#0x80
0000aaaa`bb6f72d8 97ff6c96 bl         bash+0x32530 (0000aaaa`bb6d2530)
0000aaaa`bb6f72dc aa0003fa mov        x26,x0
0000aaaa`bb6f72e0 2a1903e0 mov        w0,w25
0000aaaa`bb6f72e4 2a1a03f9 mov        w25,w26
0000aaaa`bb6f72e8 97ff6b92 bl         bash+0x32130 (0000aaaa`bb6d2130)
0000aaaa`bb6f72ec 7100035f cmp        w26,#0
```

```
0000aaaa`bb6f72f0 54fff96d ble          bash!shell_execve+0xd8 (0000aaaa`bb6f721c)  Branch

bash!shell_execve+0x1b0:
0000aaaa`bb6f72f4 51000740 sub          w0,w26,#1
0000aaaa`bb6f72f8 3820cb1f strb         wzr,[x24,w0 sxtw #0]
0000aaaa`bb6f72fc 71000b5f cmp          w26,#2
0000aaaa`bb6f7300 54fff8ed ble          bash!shell_execve+0xd8 (0000aaaa`bb6f721c)  Branch

bash!shell_execve+0x1c0:
0000aaaa`bb6f7304 394163e0 ldrb         w0,[sp,#0x58]
0000aaaa`bb6f7308 71008c1f cmp          w0,#0x23
0000aaaa`bb6f730c 54fff881 bne          bash!shell_execve+0xd8 (0000aaaa`bb6f721c)  Branch

bash!shell_execve+0x1cc:
0000aaaa`bb6f7310 394167e0 ldrb         w0,[sp,#0x59]
0000aaaa`bb6f7314 7100841f cmp          w0,#0x21
0000aaaa`bb6f7318 54fff821 bne          bash!shell_execve+0xd8 (0000aaaa`bb6f721c)  Branch

bash!shell_execve+0x1d8:
0000aaaa`bb6f731c 91016be1 add          x1,sp,#0x5A

bash!shell_execve+0x1dc:
0000aaaa`bb6f7320 39400020 ldrb         w0,[x1]
0000aaaa`bb6f7324 aa0103f7 mov          x23,x1
0000aaaa`bb6f7328 7100801f cmp          w0,#0x20
0000aaaa`bb6f732c 7a491804 ccmpne       w0,#9,#4
0000aaaa`bb6f7330 54001e81 bne          bash!shell_execve+0x5bc (0000aaaa`bb6f7700)  Branch

bash!shell_execve+0x1f0:
0000aaaa`bb6f7334 110006d6 add          w22,w22,#1
0000aaaa`bb6f7338 910006e1 add          x1,x23,#1
0000aaaa`bb6f733c 6b16033f cmp          w25,w22
0000aaaa`bb6f7340 54ffff01 bne          bash!shell_execve+0x1dc (0000aaaa`bb6f7320)  Branch

bash!shell_execve+0x200:
0000aaaa`bb6f7344 8b3ac317 add          x23,x24,w26,sxtw #0
0000aaaa`bb6f7348 d2800039 mov          x25,#1
0000aaaa`bb6f734c d280001a mov          x26,#0

bash!shell_execve+0x20c:
0000aaaa`bb6f7350 aa1903e0 mov          x0,x25
0000aaaa`bb6f7354 97ff6ad3 bl           bash+0x31ea0 (0000aaaa`bb6d1ea0)
0000aaaa`bb6f7358 aa0003f8 mov          x24,x0
0000aaaa`bb6f735c b4002720 cbz          x0,bash!shell_execve+0x6fc (0000aaaa`bb6f7840)  Branch

bash!shell_execve+0x21c:
0000aaaa`bb6f7360 aa1703e1 mov          x1,x23
0000aaaa`bb6f7364 aa1a03e2 mov          x2,x26
0000aaaa`bb6f7368 aa1803e0 mov          x0,x24
0000aaaa`bb6f736c 97ff69cd bl           bash+0x31aa0 (0000aaaa`bb6d1aa0)
0000aaaa`bb6f7370 383a6b1f strb         wzr,[x24,x26]
0000aaaa`bb6f7374 aa1803e0 mov          x0,x24
0000aaaa`bb6f7378 97ff69e6 bl           bash+0x31b10 (0000aaaa`bb6d1b10)
0000aaaa`bb6f737c b9000295 str          w21,[x20]
0000aaaa`bb6f7380 93407c15 sxtw         x21,w0
0000aaaa`bb6f7384 d10006b6 sub          x22,x21,#1
0000aaaa`bb6f7388 38766b01 ldrb         w1,[x24,x22]
0000aaaa`bb6f738c 7100343f cmp          w1,#0xD
0000aaaa`bb6f7390 54001ea0 beq          bash!shell_execve+0x620 (0000aaaa`bb6f7764)  Branch

bash!shell_execve+0x250:
0000aaaa`bb6f7394 528000a2 mov          w2,#5
0000aaaa`bb6f7398 d0000641 adrp         x1,bash!IO_stdin_used+0x11738 (0000aaaa`bb7c1000)
0000aaaa`bb6f739c d2800000 mov          x0,#0
0000aaaa`bb6f73a0 91324021 add          x1,x1,#0xC90
0000aaaa`bb6f73a4 97ff6ca3 bl           bash+0x32630 (0000aaaa`bb6d2630)
0000aaaa`bb6f73a8 52800fd9 mov          w25,#0x7E
0000aaaa`bb6f73ac aa1303e1 mov          x1,x19
0000aaaa`bb6f73b0 aa1803e2 mov          x2,x24
0000aaaa`bb6f73b4 94003b44 bl           bash!sys_error (0000aaaa`bb7060c4)
0000aaaa`bb6f73b8 aa1803e0 mov          x0,x24
0000aaaa`bb6f73bc 97ff6bd9 bl           bash+0x32320 (0000aaaa`bb6d2320)
0000aaaa`bb6f73c0 17ffffa1 b            bash!shell_execve+0x100 (0000aaaa`bb6f7244)  Branch

bash!shell_execve+0x280:
0000aaaa`bb6f73c4 aa1303e0 mov          x0,x19
```

88

```
0000aaaa`bb6f73c8 52800001 mov          w1,#0
0000aaaa`bb6f73cc 97ff6abd bl           bash+0x31ec0 (0000aaaa`bb6d1ec0)
0000aaaa`bb6f73d0 2a0003f9 mov          w25,w0
0000aaaa`bb6f73d4 36f81480 tbz          x0,#0x1F,bash!shell_execve+0x520 (0000aaaa`bb6f7664)  Branch


bash!shell_execve+0x294:
0000aaaa`bb6f73d8 b0000854 adrp         x20,bash!o_options+0x2118 (0000aaaa`bb800000)
0000aaaa`bb6f73dc 97ff77dd bl           bash!reset_parser (0000aaaa`bb6d5350)
0000aaaa`bb6f73e0 f9421a80 ldr          x0,[x20,#0x430]
0000aaaa`bb6f73e4 f9400000 ldr          x0,[x0]
0000aaaa`bb6f73e8 b4000240 cbz          x0,bash!shell_execve+0x2ec (0000aaaa`bb6f7430)  Branch


bash!shell_execve+0x2a8:
0000aaaa`bb6f73ec b9400c01 ldr          w1,[x0,#0xC]
0000aaaa`bb6f73f0 34000081 cbz          w1,bash!shell_execve+0x2bc (0000aaaa`bb6f7400)  Branch


bash!shell_execve+0x2b0:
0000aaaa`bb6f73f4 f0000181 adrp         x1,bash!shell_glob_filename+0x10e0 (0000aaaa`bb72a000)
0000aaaa`bb6f73f8 9115c021 add          x1,x1,#0x570
0000aaaa`bb6f73fc 9400a5d2 bl           bash!phash_flush+0xb4 (0000aaaa`bb720b44)


bash!shell_execve+0x2bc:
0000aaaa`bb6f7400 f9421a94 ldr          x20,[x20,#0x430]
0000aaaa`bb6f7404 f9400295 ldr          x21,[x20]
0000aaaa`bb6f7408 f94002a0 ldr          x0,[x21]
0000aaaa`bb6f740c 97ff6bc5 bl           bash+0x32320 (0000aaaa`bb6d2320)
0000aaaa`bb6f7410 aa1503e0 mov          x0,x21
0000aaaa`bb6f7414 97ff6bc3 bl           bash+0x32320 (0000aaaa`bb6d2320)
0000aaaa`bb6f7418 b0000840 adrp         x0,bash!o_options+0x2118 (0000aaaa`bb800000)
0000aaaa`bb6f741c f900029f str          xzr,[x20]
0000aaaa`bb6f7420 f942d400 ldr          x0,[x0,#0x5A8]
0000aaaa`bb6f7424 b9400001 ldr          w1,[x0]
0000aaaa`bb6f7428 321f0021 orr          w1,w1,#2
0000aaaa`bb6f742c b9000001 str          w1,[x0]


bash!shell_execve+0x2ec:
0000aaaa`bb6f7430 b0000840 adrp         x0,bash!o_options+0x2118 (0000aaaa`bb800000)
0000aaaa`bb6f7434 f9407c00 ldr          x0,[x0,#0xF8]
0000aaaa`bb6f7438 b900001f str          wzr,[x0]
0000aaaa`bb6f743c 9400661d bl           bash!without_job_control (0000aaaa`bb710cb0)
0000aaaa`bb6f7440 52800220 mov          w0,#0x11
0000aaaa`bb6f7444 f0000041 adrp         x1,bash!initialize_shell_variables+0x26bc (0000aaaa`bb702000)
0000aaaa`bb6f7448 91040021 add          x1,x1,#0x100
0000aaaa`bb6f744c 9400ccfe bl           bash!set_signal_handler (0000aaaa`bb72a844)
0000aaaa`bb6f7450 f0000641 adrp         x1,bash!is_basic_table+0x490 (0000aaaa`bb7c2000)
0000aaaa`bb6f7454 91144021 add          x1,x1,#0x510
0000aaaa`bb6f7458 90000840 adrp         x0,bash!o_options+0x1118 (0000aaaa`bb7ff000)
0000aaaa`bb6f745c f9465c00 ldr          x0,[x0,#0xCB8]
0000aaaa`bb6f7460 ad400c22 ldp          q2,q3,[x1]
0000aaaa`bb6f7464 ad410420 ldp          q0,q1,[x1,#0x20]
0000aaaa`bb6f7468 f9402021 ldr          x1,[x1,#0x40]
0000aaaa`bb6f746c ad000c02 stp          q2,q3,[x0]
0000aaaa`bb6f7470 ad010400 stp          q0,q1,[x0,#0x20]
0000aaaa`bb6f7474 f9002001 str          x1,[x0,#0x40]
0000aaaa`bb6f7478 94002bc6 bl           bash!reset_shell_flags (0000aaaa`bb702390)
0000aaaa`bb6f747c b0000845 adrp         x5,bash!o_options+0x2118 (0000aaaa`bb800000)
0000aaaa`bb6f7480 90000844 adrp         x4,bash!o_options+0x1118 (0000aaaa`bb7ff000)
0000aaaa`bb6f7484 b0000843 adrp         x3,bash!o_options+0x2118 (0000aaaa`bb800000)
0000aaaa`bb6f7488 b0000842 adrp         x2,bash!o_options+0x2118 (0000aaaa`bb800000)
0000aaaa`bb6f748c b0000841 adrp         x1,bash!o_options+0x2118 (0000aaaa`bb800000)
0000aaaa`bb6f7490 b0000840 adrp         x0,bash!o_options+0x2118 (0000aaaa`bb800000)
0000aaaa`bb6f7494 f9456842 ldr          x2,[x2,#0xAD0]
0000aaaa`bb6f7498 52800026 mov          w6,#1
0000aaaa`bb6f749c f943f421 ldr          x1,[x1,#0x7E8]
0000aaaa`bb6f74a0 f9446800 ldr          x0,[x0,#0x8D0]
0000aaaa`bb6f74a4 b900005f str          wzr,[x2]
0000aaaa`bb6f74a8 f94164a5 ldr          x5,[x5,#0x2C8]
0000aaaa`bb6f74ac b9000026 str          w6,[x1]
0000aaaa`bb6f74b0 f9467484 ldr          x4,[x4,#0xCE8]
0000aaaa`bb6f74b4 b9000006 str          w6,[x0]
0000aaaa`bb6f74b8 f9441063 ldr          x3,[x3,#0x820]
0000aaaa`bb6f74bc b90000bf str          wzr,[x5]
0000aaaa`bb6f74c0 b900009f str          wzr,[x4]
0000aaaa`bb6f74c4 b900007f str          wzr,[x3]
0000aaaa`bb6f74c8 9401a546 bl           bash!reset_shopt_options (0000aaaa`bb7609e0)
0000aaaa`bb6f74cc b0000840 adrp         x0,bash!o_options+0x2118 (0000aaaa`bb800000)
```

```
0000aaaa`bb6f74d0 f9456c00 ldr        x0,[x0,#0xAD8]
0000aaaa`bb6f74d4 f9400001 ldr        x1,[x0]
0000aaaa`bb6f74d8 b9400c22 ldr        w2,[x1,#0xC]
0000aaaa`bb6f74dc 36180062 tbz        x2,#3,bash!shell_execve+0x3a4 (0000aaaa`bb6f74e8)   Branch

bash!shell_execve+0x39c:
0000aaaa`bb6f74e0 f9400c21 ldr        x1,[x1,#0x18]
0000aaaa`bb6f74e4 f9000001 str        x1,[x0]

bash!shell_execve+0x3a4:
0000aaaa`bb6f74e8 f0000880 adrp       x0,bash!subshell_top_level+0x50 (0000aaaa`bb80a000)
0000aaaa`bb6f74ec f9415401 ldr        x1,[x0,#0x2A8]
0000aaaa`bb6f74f0 b4000041 cbz        x1,bash!shell_execve+0x3b4 (0000aaaa`bb6f74f8)   Branch

bash!shell_execve+0x3b0:
0000aaaa`bb6f74f4 f901541f str        xzr,[x0,#0x2A8]

bash!shell_execve+0x3b4:
0000aaaa`bb6f74f8 b0000842 adrp       x2,bash!o_options+0x2118 (0000aaaa`bb800000)
0000aaaa`bb6f74fc b0000847 adrp       x7,bash!o_options+0x2118 (0000aaaa`bb800000)
0000aaaa`bb6f7500 b0000846 adrp       x6,bash!o_options+0x2118 (0000aaaa`bb800000)
0000aaaa`bb6f7504 b0000845 adrp       x5,bash!o_options+0x2118 (0000aaaa`bb800000)
0000aaaa`bb6f7508 f9431042 ldr        x2,[x2,#0x620]
0000aaaa`bb6f750c b0000844 adrp       x4,bash!o_options+0x2118 (0000aaaa`bb800000)
0000aaaa`bb6f7510 b0000843 adrp       x3,bash!o_options+0x2118 (0000aaaa`bb800000)
0000aaaa`bb6f7514 b0000841 adrp       x1,bash!o_options+0x2118 (0000aaaa`bb800000)
0000aaaa`bb6f7518 b0000840 adrp       x0,bash!o_options+0x2118 (0000aaaa`bb800000)
0000aaaa`bb6f751c f945d0e7 ldr        x7,[x7,#0xBA0]
0000aaaa`bb6f7520 f94048c6 ldr        x6,[x6,#0x90]
0000aaaa`bb6f7524 f946b0a5 ldr        x5,[x5,#0xD60]
0000aaaa`bb6f7528 b90000ff str        wzr,[x7]
0000aaaa`bb6f752c f9423484 ldr        x4,[x4,#0x468]
0000aaaa`bb6f7530 b90000df str        wzr,[x6]
0000aaaa`bb6f7534 f9450063 ldr        x3,[x3,#0xA00]
0000aaaa`bb6f7538 b90000bf str        wzr,[x5]
0000aaaa`bb6f753c f941a421 ldr        x1,[x1,#0x348]
0000aaaa`bb6f7540 b900009f str        wzr,[x4]
0000aaaa`bb6f7544 f946a800 ldr        x0,[x0,#0xD50]
0000aaaa`bb6f7548 b900007f str        wzr,[x3]
0000aaaa`bb6f754c b9400042 ldr        w2,[x2]
0000aaaa`bb6f7550 b900003f str        wzr,[x1]
0000aaaa`bb6f7554 b900001f str        wzr,[x0]
0000aaaa`bb6f7558 34001542 cbz        w2,bash!shell_execve+0x6bc (0000aaaa`bb6f7800)   Branch

bash!shell_execve+0x418:
0000aaaa`bb6f755c 9400be09 bl         bash!set_sigint_handler (0000aaaa`bb726d80)
0000aaaa`bb6f7560 aa1603e0 mov        x0,x22
0000aaaa`bb6f7564 9401da13 bl         bash!strvec_len (0000aaaa`bb76ddb0)
0000aaaa`bb6f7568 2a0003f5 mov        w21,w0
0000aaaa`bb6f756c aa1603e0 mov        x0,x22
0000aaaa`bb6f7570 11000aa1 add        w1,w21,#2
0000aaaa`bb6f7574 9401d937 bl         bash!strvec_resize (0000aaaa`bb76da50)
0000aaaa`bb6f7578 93407eb6 sxtw       x22,w21
0000aaaa`bb6f757c aa0003f4 mov        x20,x0
0000aaaa`bb6f7580 910006d6 add        x22,x22,#1
0000aaaa`bb6f7584 d37d7ea2 ubfiz      x2,x21,#3,#0x20
0000aaaa`bb6f7588 cb3542c0 sub        x0,x22,w21,uxtw #0
0000aaaa`bb6f758c 110006b5 add        w21,w21,#1
0000aaaa`bb6f7590 d37df000 lsl        x0,x0,#3
0000aaaa`bb6f7594 d1002001 sub        x1,x0,#8
0000aaaa`bb6f7598 8b000280 add        x0,x20,x0
0000aaaa`bb6f759c 8b010281 add        x1,x20,x1
0000aaaa`bb6f75a0 97ff6948 bl         bash+0x31ac0 (0000aaaa`bb6d1ac0)
0000aaaa`bb6f75a4 b0000840 adrp       x0,bash!o_options+0x2118 (0000aaaa`bb800000)
0000aaaa`bb6f75a8 f9413c00 ldr        x0,[x0,#0x278]
0000aaaa`bb6f75ac f9400000 ldr        x0,[x0]
0000aaaa`bb6f75b0 a9004e80 stp        x0,x19,[x20]
0000aaaa`bb6f75b4 f8367a9f str        xzr,[x20,x22 lsl #3]
0000aaaa`bb6f75b8 f9400280 ldr        x0,[x20]
0000aaaa`bb6f75bc 39400001 ldrb       w1,[x0]
0000aaaa`bb6f75c0 7100b43f cmp        w1,#0x2D
0000aaaa`bb6f75c4 54000061 bne        bash!shell_execve+0x48c (0000aaaa`bb6f75d0)   Branch

bash!shell_execve+0x484:
0000aaaa`bb6f75c8 91000400 add        x0,x0,#1
0000aaaa`bb6f75cc f9000280 str        x0,[x20]
```

```
bash!shell_execve+0x48c:
0000aaaa`bb6f75d0 90000840 adrp        x0,bash!o_options+0x1118 (0000aaaa`bb7ff000)
0000aaaa`bb6f75d4 f947e000 ldr         x0,[x0,#0xFC0]
0000aaaa`bb6f75d8 b9400000 ldr         w0,[x0]
0000aaaa`bb6f75dc 350010a0 cbnz        w0,bash!shell_execve+0x6ac (0000aaaa`bb6f77f0)   Branch

bash!shell_execve+0x49c:
0000aaaa`bb6f75e0 b0000856 adrp        x22,bash!o_options+0x2118 (0000aaaa`bb800000)
0000aaaa`bb6f75e4 b0000853 adrp        x19,bash!o_options+0x2118 (0000aaaa`bb800000)
0000aaaa`bb6f75e8 f94102d9 ldr         x25,[x22,#0x200]
0000aaaa`bb6f75ec f9400320 ldr         x0,[x25]
0000aaaa`bb6f75f0 b4000180 cbz         x0,bash!shell_execve+0x4dc (0000aaaa`bb6f7620)   Branch

bash!shell_execve+0x4b0:
0000aaaa`bb6f75f4 f946567a ldr         x26,[x19,#0xCA8]
0000aaaa`bb6f75f8 d2800018 mov         x24,#0
0000aaaa`bb6f75fc 14000003 b           bash!shell_execve+0x4c4 (0000aaaa`bb6f7608)   Branch

bash!shell_execve+0x4bc:
0000aaaa`bb6f7600 f8787800 ldr         x0,[x0,x24 lsl #3]
0000aaaa`bb6f7604 97ff6b47 bl          bash+0x32320 (0000aaaa`bb6d2320)

bash!shell_execve+0x4c4:
0000aaaa`bb6f7608 b9400341 ldr         w1,[x26]
0000aaaa`bb6f760c 91000718 add         x24,x24,#1
0000aaaa`bb6f7610 f9400320 ldr         x0,[x25]
0000aaaa`bb6f7614 6b18003f cmp         w1,w24
0000aaaa`bb6f7618 54ffff4c bgt         bash!shell_execve+0x4bc (0000aaaa`bb6f7600)   Branch

bash!shell_execve+0x4d8:
0000aaaa`bb6f761c 97ff6b41 bl          bash+0x32320 (0000aaaa`bb6d2320)

bash!shell_execve+0x4dc:
0000aaaa`bb6f7620 f0000898 adrp        x24,bash!subshell_top_level+0x50 (0000aaaa`bb80a000)
0000aaaa`bb6f7624 f9413f00 ldr         x0,[x24,#0x278]
0000aaaa`bb6f7628 97ffc472 bl          bash!dispose_command (0000aaaa`bb6e87f0)
0000aaaa`bb6f762c f9013f1f str         xzr,[x24,#0x278]
0000aaaa`bb6f7630 90000840 adrp        x0,bash!o_options+0x1118 (0000aaaa`bb7ff000)
0000aaaa`bb6f7634 f9465673 ldr         x19,[x19,#0xCA8]
0000aaaa`bb6f7638 f9462000 ldr         x0,[x0,#0xC40]
0000aaaa`bb6f763c f94102d6 ldr         x22,[x22,#0x200]
0000aaaa`bb6f7640 b9000275 str         w21,[x19]
0000aaaa`bb6f7644 f9000017 str         x23,[x0]
0000aaaa`bb6f7648 f90002d4 str         x20,[x22]
0000aaaa`bb6f764c 97ff7a5d bl          bash!unbind_args (0000aaaa`bb6d5fc0)
0000aaaa`bb6f7650 94008040 bl          bash!clear_fifo_list (0000aaaa`bb717750)
0000aaaa`bb6f7654 90000840 adrp        x0,bash!o_options+0x1118 (0000aaaa`bb7ff000)
0000aaaa`bb6f7658 52800021 mov         w1,#1
0000aaaa`bb6f765c f947ec00 ldr         x0,[x0,#0xFD8]
0000aaaa`bb6f7660 97ff6b78 bl          bash+0x32440 (0000aaaa`bb6d2440)

bash!shell_execve+0x520:
0000aaaa`bb6f7664 910163f8 add         x24,sp,#0x58
0000aaaa`bb6f7668 d2801002 mov         x2,#0x80
0000aaaa`bb6f766c aa1803e1 mov         x1,x24
0000aaaa`bb6f7670 97ff6bb0 bl          bash+0x32530 (0000aaaa`bb6d2530)
0000aaaa`bb6f7674 aa0003f5 mov         x21,x0
0000aaaa`bb6f7678 2a1903e0 mov         w0,w25
0000aaaa`bb6f767c 2a1503f9 mov         w25,w21
0000aaaa`bb6f7680 97ff6aac bl          bash+0x32130 (0000aaaa`bb6d2130)
0000aaaa`bb6f7684 710002bf cmp         w21,#0
0000aaaa`bb6f7688 34ffddf5 cbz         w21,bash!shell_execve+0x100 (0000aaaa`bb6f7244)   Branch

bash!shell_execve+0x548:
0000aaaa`bb6f768c 54ffea6d ble         bash!shell_execve+0x294 (0000aaaa`bb6f73d8)   Branch

bash!shell_execve+0x54c:
0000aaaa`bb6f7690 510006a1 sub         w1,w21,#1
0000aaaa`bb6f7694 d2800023 mov         x3,#1
0000aaaa`bb6f7698 91000821 add         x1,x1,#2
0000aaaa`bb6f769c 14000004 b           bash!shell_execve+0x568 (0000aaaa`bb6f76ac)   Branch

bash!shell_execve+0x55c:
0000aaaa`bb6f76a0 91000463 add         x3,x3,#1
0000aaaa`bb6f76a4 eb01007f cmp         x3,x1
```

```
0000aaaa`bb6f76a8 54ffe980 beq             bash!shell_execve+0x294 (0000aaaa`bb6f73d8)   Branch

bash!shell_execve+0x568:
0000aaaa`bb6f76ac 8b030300 add             x0,x24,x3
0000aaaa`bb6f76b0 385ff000 ldurb           w0,[x0,#-1]
0000aaaa`bb6f76b4 7100281f cmp             w0,#0xA
0000aaaa`bb6f76b8 54ffe900 beq             bash!shell_execve+0x294 (0000aaaa`bb6f73d8)   Branch

bash!shell_execve+0x578:
0000aaaa`bb6f76bc 35ffff20 cbnz            w0,bash!shell_execve+0x55c (0000aaaa`bb6f76a0)   Branch

bash!shell_execve+0x57c:
0000aaaa`bb6f76c0 d0000641 adrp            x1,bash!IO_stdin_used+0x11738 (0000aaaa`bb7c1000)
0000aaaa`bb6f76c4 9132a021 add             x1,x1,#0xCA8
0000aaaa`bb6f76c8 528000a2 mov             w2,#5
0000aaaa`bb6f76cc d2800000 mov             x0,#0
0000aaaa`bb6f76d0 97ff6bd8 bl              bash+0x32630 (0000aaaa`bb6d2630)
0000aaaa`bb6f76d4 aa0003f5 mov             x21,x0
0000aaaa`bb6f76d8 52800100 mov             w0,#8
0000aaaa`bb6f76dc 97ff6a8d bl              bash+0x32110 (0000aaaa`bb6d2110)
0000aaaa`bb6f76e0 aa1303e1 mov             x1,x19
0000aaaa`bb6f76e4 aa0003e2 mov             x2,x0
0000aaaa`bb6f76e8 aa1503e0 mov             x0,x21
0000aaaa`bb6f76ec 940038d1 bl              bash!internal_error (0000aaaa`bb705a30)
0000aaaa`bb6f76f0 52800100 mov             w0,#8
0000aaaa`bb6f76f4 52800fd9 mov             w25,#0x7E
0000aaaa`bb6f76f8 b9000280 str             w0,[x20]
0000aaaa`bb6f76fc 17fffed2 b               bash!shell_execve+0x100 (0000aaaa`bb6f7244)   Branch

bash!shell_execve+0x5bc:
0000aaaa`bb6f7700 6b16035f cmp             w26,w22
0000aaaa`bb6f7704 5400098d ble             bash!shell_execve+0x6f0 (0000aaaa`bb6f7834)   Branch

bash!shell_execve+0x5c4:
0000aaaa`bb6f7708 110006c3 add             w3,w22,#1
0000aaaa`bb6f770c 2a1603e2 mov             w2,w22
0000aaaa`bb6f7710 8b23c303 add             x3,x24,w3,sxtw #0
0000aaaa`bb6f7714 14000005 b               bash!shell_execve+0x5e4 (0000aaaa`bb6f7728)   Branch

bash!shell_execve+0x5d4:
0000aaaa`bb6f7718 11000442 add             w2,w2,#1
0000aaaa`bb6f771c 6b02033f cmp             w25,w2
0000aaaa`bb6f7720 54000180 beq             bash!shell_execve+0x60c (0000aaaa`bb6f7750)   Branch

bash!shell_execve+0x5e0:
0000aaaa`bb6f7724 38401460 ldrb            w0,[x3],#1

bash!shell_execve+0x5e4:
0000aaaa`bb6f7728 51002401 sub             w1,w0,#9
0000aaaa`bb6f772c 7100801f cmp             w0,#0x20
0000aaaa`bb6f7730 12001c20 and             w0,w1,#0xFF
0000aaaa`bb6f7734 7a411800 ccmpne          w0,#1,#0
0000aaaa`bb6f7738 54ffff08 bhi             bash!shell_execve+0x5d4 (0000aaaa`bb6f7718)   Branch

bash!shell_execve+0x5f8:
0000aaaa`bb6f773c 4b160042 sub             w2,w2,w22
0000aaaa`bb6f7740 11000459 add             w25,w2,#1
0000aaaa`bb6f7744 93407c5a sxtw            x26,w2
0000aaaa`bb6f7748 93407f39 sxtw            x25,w25
0000aaaa`bb6f774c 17ffff01 b               bash!shell_execve+0x20c (0000aaaa`bb6f7350)   Branch

bash!shell_execve+0x60c:
0000aaaa`bb6f7750 4b16035a sub             w26,w26,w22
0000aaaa`bb6f7754 11000759 add             w25,w26,#1
0000aaaa`bb6f7758 93407f5a sxtw            x26,w26
0000aaaa`bb6f775c 93407f39 sxtw            x25,w25
0000aaaa`bb6f7760 17fffefc b               bash!shell_execve+0x20c (0000aaaa`bb6f7350)   Branch

bash!shell_execve+0x620:
0000aaaa`bb6f7764 11000814 add             w20,w0,#2
0000aaaa`bb6f7768 aa1803e0 mov             x0,x24
0000aaaa`bb6f776c 93407e94 sxtw            x20,w20
0000aaaa`bb6f7770 aa1403e1 mov             x1,x20
0000aaaa`bb6f7774 97ff6a4f bl              bash+0x320b0 (0000aaaa`bb6d20b0)
0000aaaa`bb6f7778 910006a1 add             x1,x21,#1
0000aaaa`bb6f777c aa0003f8 mov             x24,x0
```

```
0000aaaa`bb6f7780 b40000e0 cbz        x0,bash!shell_execve+0x658 (0000aaaa`bb6f779c)  Branch

bash!shell_execve+0x640:
0000aaaa`bb6f7784 52800bc0 mov        w0,#0x5E
0000aaaa`bb6f7788 38366b00 strb       w0,[x24,x22]
0000aaaa`bb6f778c 528009a0 mov        w0,#0x4D
0000aaaa`bb6f7790 38356b00 strb       w0,[x24,x21]
0000aaaa`bb6f7794 38216b1f strb       wzr,[x24,x1]
0000aaaa`bb6f7798 17fffeff b          bash!shell_execve+0x250 (0000aaaa`bb6f7394)  Branch

bash!shell_execve+0x658:
0000aaaa`bb6f779c aa1403e1 mov        x1,x20
0000aaaa`bb6f77a0 900005c0 adrp       x0,bash!rl_enable_paren_matching+0x87d0 (0000aaaa`bb7af000)
0000aaaa`bb6f77a4 91332000 add        x0,x0,#0xCC8
0000aaaa`bb6f77a8 94012bca bl         bash!setup_exec_ignore+0x10 (0000aaaa`bb7426d0)
0000aaaa`bb6f77ac 52800bc1 mov        w1,#0x5E
0000aaaa`bb6f77b0 528009a0 mov        w0,#0x4D
0000aaaa`bb6f77b4 381ff2a1 sturb      w1,[x21,#-1]
0000aaaa`bb6f77b8 528000a2 mov        w2,#5
0000aaaa`bb6f77bc 390002a0 strb       w0,[x21]
0000aaaa`bb6f77c0 d0000641 adrp       x1,bash!IO_stdin_used+0x11738 (0000aaaa`bb7c1000)
0000aaaa`bb6f77c4 390006bf strb       wzr,[x21,#1]
0000aaaa`bb6f77c8 91324021 add        x1,x1,#0xC90
0000aaaa`bb6f77cc d2800000 mov        x0,#0
0000aaaa`bb6f77d0 52800fd9 mov        w25,#0x7E
0000aaaa`bb6f77d4 97ff6b97 bl         bash+0x32630 (0000aaaa`bb6d2630)
0000aaaa`bb6f77d8 aa1303e1 mov        x1,x19
0000aaaa`bb6f77dc d0000642 adrp       x2,bash!IO_stdin_used+0x11738 (0000aaaa`bb7c1000)
0000aaaa`bb6f77e0 91236042 add        x2,x2,#0x8D8
0000aaaa`bb6f77e4 94003a38 bl         bash!sys_error (0000aaaa`bb7060c4)
0000aaaa`bb6f77e8 17fffe97 b          bash!shell_execve+0x100 (0000aaaa`bb6f7244)  Branch

bash!shell_execve+0x6a8:
0000aaaa`bb6f77ec 97ff6a4d bl         bash+0x32120 (0000aaaa`bb6d2120)

bash!shell_execve+0x6ac:
0000aaaa`bb6f77f0 52800561 mov        w1,#0x2B
0000aaaa`bb6f77f4 52800e40 mov        w0,#0x72
0000aaaa`bb6f77f8 94004516 bl         bash!change_flag (0000aaaa`bb708c50)
0000aaaa`bb6f77fc 17ffff79 b          bash!shell_execve+0x49c (0000aaaa`bb6f75e0)  Branch

bash!shell_execve+0x6bc:
0000aaaa`bb6f7800 90000854 adrp       x20,bash!o_options+0x1118 (0000aaaa`bb7ff000)
0000aaaa`bb6f7804 f946ae94 ldr        x20,[x20,#0xD58]
0000aaaa`bb6f7808 b9400280 ldr        w0,[x20]
0000aaaa`bb6f780c 7100001f cmp        w0,#0
0000aaaa`bb6f7810 54ffea6d ble        bash!shell_execve+0x418 (0000aaaa`bb6f755c)  Branch

bash!shell_execve+0x6d0:
0000aaaa`bb6f7814 9400ab13 bl         bash!close_buffered_fd (0000aaaa`bb722460)
0000aaaa`bb6f7818 b0000840 adrp       x0,bash!o_options+0x2118 (0000aaaa`bb800000)
0000aaaa`bb6f781c 12800001 mov        w1,#-1
0000aaaa`bb6f7820 b9000281 str        w1,[x20]
0000aaaa`bb6f7824 f945a400 ldr        x0,[x0,#0xB48]
0000aaaa`bb6f7828 b900001f str        wzr,[x0]
0000aaaa`bb6f782c b9001001 str        w1,[x0,#0x10]
0000aaaa`bb6f7830 17ffff4b b          bash!shell_execve+0x418 (0000aaaa`bb6f755c)  Branch

bash!shell_execve+0x6f0:
0000aaaa`bb6f7834 d280001a mov        x26,#0
0000aaaa`bb6f7838 d2800039 mov        x25,#1
0000aaaa`bb6f783c 17fffcc5 b          bash!shell_execve+0x20c (0000aaaa`bb6f7350)  Branch

bash!shell_execve+0x6fc:
0000aaaa`bb6f7840 aa1903e1 mov        x1,x25
0000aaaa`bb6f7844 900005c0 adrp       x0,bash!rl_enable_paren_matching+0x87d0 (0000aaaa`bb7af000)
0000aaaa`bb6f7848 91236000 add        x0,x0,#0x8D8
0000aaaa`bb6f784c 94012ba1 bl         bash!setup_exec_ignore+0x10 (0000aaaa`bb7426d0)
0000aaaa`bb6f7850 17fffec4 b          bash!shell_execve+0x21c (0000aaaa`bb6f7360)  Branch
```

93

4.   Follow the call path to see the real branch and link to the *libc.so.6* library:

```
0:000> u 0xaaaabb6d1ec0
bash+0x31ec0:
0000aaaa`bb6d1ec0 d0000970 adrp     xip0,bash!o_options+0x1118 (0000aaaa`bb7ff000)
0000aaaa`bb6d1ec4 f942de11 ldr      xip1,[xip0,#0x5B8]
0000aaaa`bb6d1ec8 9116e210 add      xip0,xip0,#0x5B8
0000aaaa`bb6d1ecc d61f0220 br       xip1
0000aaaa`bb6d1ed0 d0000970 adrp     xip0,bash!o_options+0x1118 (0000aaaa`bb7ff000)
0000aaaa`bb6d1ed4 f942e211 ldr      xip1,[xip0,#0x5C0]
0000aaaa`bb6d1ed8 91170210 add      xip0,xip0,#0x5C0
0000aaaa`bb6d1edc d61f0220 br       xip1
```

```
0:000> dps 0000aaaa`bb7ff000+0x5B8
0000aaaa`bb7ff5b8  0000ffff`bafc7810 libc_so!_open64
0000aaaa`bb7ff5c0  0000ffff`bafe57c0 libc_so!_fdelt_chk
0000aaaa`bb7ff5c8  0000ffff`bafe39c0 libc_so!_strncpy_chk
0000aaaa`bb7ff5d0  0000ffff`baf98240 libc_so!tzset
0000aaaa`bb7ff5d8  0000ffff`bafe3860 libc_so!_strcpy_chk
0000aaaa`bb7ff5e0  0000ffff`baf92970 libc_so!wcswidth
0000aaaa`bb7ff5e8  0000ffff`bafa8040 libc_so!getppid
0000aaaa`bb7ff5f0  0000ffff`baf2b040 libc_so!sigemptyset
0000aaaa`bb7ff5f8  0000ffff`baf807e4 libc_so!strncmp
0000aaaa`bb7ff600  0000ffff`baf89ce0 libc_so!wcsncmp
0000aaaa`bb7ff608  0000ffff`baf248a0 libc_so!bindtextdomain
0000aaaa`bb7ff610  0000ffff`baf2b360 libc_so!_libc_current_sigrtmin
0000aaaa`bb7ff618  0000ffff`baf7ff00 libc_so!strcat
0000aaaa`bb7ff620  0000ffff`bafe3c10 libc_so!_printf_chk
0000aaaa`bb7ff628  0000ffff`baf62fc0 libc_so!_fpurge
0000aaaa`bb7ff630  0000ffff`baf88fc0 libc_so!strerrordesc_np+0x14a0
```

5.   We can also check the regions for loaded shared libraries:

```
0:000> !address
```

	BaseAddress	EndAddress+1	RegionSize	Type	State	Protect	Usage	
+	0`00000000	aaaa`bb6a0000	aaaa`bb6a0000				\<unknown\>	
+	aaaa`bb6a0000	aaaa`bb7ed000	0`0014d000	MEM_PRIVATE	MEM_COMMIT	PAGE_EXECUTE_READ	Image	[bash; "/usr/bin/bash"]
+	aaaa`bb7ed000	aaaa`bb7fc000	0`0000f000				Image	[bash; "/usr/bin/bash"]
+	aaaa`bb7fc000	aaaa`bb801000	0`00005000	MEM_PRIVATE	MEM_COMMIT	PAGE_READONLY	Image	[bash; "/usr/bin/bash"]
+	aaaa`bb801000	aaaa`bb80a000	0`00009000	MEM_PRIVATE	MEM_COMMIT	PAGE_READWRITE	Image	[bash; "/usr/bin/bash"]
+	aaaa`bb80a000	aaaa`bb815000	0`0000b000	MEM_PRIVATE	MEM_COMMIT	PAGE_READWRITE	\<unknown\>	[...........Z...]
+	aaaa`bb815000	aaaa`e4664000	0`28e4f000				\<unknown\>	
+	aaaa`e4664000	aaaa`e4815000	0`001b1000	MEM_PRIVATE	MEM_COMMIT	PAGE_READWRITE	\<unknown\>	[...............]
+	aaaa`e4815000	ffff`babb0000	5554`d639b000				\<unknown\>	
+	ffff`babb0000	ffff`bac07000	0`00057000	MEM_PRIVATE	MEM_COMMIT	PAGE_EXECUTE_READ	Image	[LC_CTYPE; "/usr/lib/locale/C.utf8/LC_CTYPE"]
+	ffff`bac07000	ffff`baef0000	0`002e9000	MEM_PRIVATE	MEM_COMMIT	PAGE_EXECUTE_READ	Image	[locale_archive; "/usr/lib/locale/locale-archive"]
+	ffff`baef0000	ffff`bb079000	0`00189000	MEM_PRIVATE	MEM_COMMIT	PAGE_EXECUTE_READ	Image	[libc_so; "/usr/lib/aarch64-linux-gnu/libc.so.6"]
+	ffff`bb079000	ffff`bb088000	0`0000f000				Image	[libc_so; "/usr/lib/aarch64-linux-gnu/libc.so.6"]
+	ffff`bb088000	ffff`bb08c000	0`00004000	MEM_PRIVATE	MEM_COMMIT	PAGE_READONLY	Image	[libc_so; "/usr/lib/aarch64-linux-gnu/libc.so.6"]
+	ffff`bb08c000	ffff`bb08e000	0`00002000	MEM_PRIVATE	MEM_COMMIT	PAGE_READWRITE	Image	[libc_so; "/usr/lib/aarch64-linux-gnu/libc.so.6"]
+	ffff`bb08e000	ffff`bb09a000	0`0000c000	MEM_PRIVATE	MEM_COMMIT	PAGE_READWRITE	\<unknown\>	[................]
+	ffff`bb09a000	ffff`bb0a0000	0`00006000				\<unknown\>	
+	ffff`bb0a0000	ffff`bb0cc000	0`0002c000	MEM_PRIVATE	MEM_COMMIT	PAGE_EXECUTE_READ	Image	[libtinfo_so_6; "/usr/lib/aarch64-linux-gnu/libtinfo.so.6.3"]
+	ffff`bb0cc000	ffff`bb0db000	0`0000f000				Image	[libtinfo_so_6; "/usr/lib/aarch64-linux-gnu/libtinfo.so.6.3"]
+	ffff`bb0db000	ffff`bb0df000	0`00004000	MEM_PRIVATE	MEM_COMMIT	PAGE_READONLY	Image	[libtinfo_so_6; "/usr/lib/aarch64-linux-gnu/libtinfo.so.6.3"]
+	ffff`bb0df000	ffff`bb0e0000	0`00001000	MEM_PRIVATE	MEM_COMMIT	PAGE_READWRITE	Image	[libtinfo_so_6; "/usr/lib/aarch64-linux-gnu/libtinfo.so.6.3"]
+	ffff`bb0e0000	ffff`bb0e3000	0`00003000				\<unknown\>	
+	ffff`bb0e3000	ffff`bb0e4000	0`00001000	MEM_PRIVATE	MEM_COMMIT	PAGE_EXECUTE_READ	Image	[LC_NUMERIC; "/usr/lib/locale/C.utf8/LC_NUMERIC"]
+	ffff`bb0e4000	ffff`bb0e5000	0`00001000	MEM_PRIVATE	MEM_COMMIT	PAGE_EXECUTE_READ	Image	[LC_TIME; "/usr/lib/locale/C.utf8/LC_TIME"]
+	ffff`bb0e5000	ffff`bb0e6000	0`00001000	MEM_PRIVATE	MEM_COMMIT	PAGE_EXECUTE_READ	Image	[LC_COLLATE; "/usr/lib/locale/C.utf8/LC_COLLATE"]
+	ffff`bb0e6000	ffff`bb0ed000	0`00007000	MEM_PRIVATE	MEM_COMMIT	PAGE_EXECUTE_READ	Image	[gconv_modules; "/usr/lib/aarch64-linux-gnu/gconv/gconv-modules.cache"]
+	ffff`bb0ed000	ffff`bb118000	0`0002b000	MEM_PRIVATE	MEM_COMMIT	PAGE_EXECUTE_READ	Image	[ld_linux_aarch64_so; "/usr/lib/aarch64-linux-gnu/ld-linux-aarch64.so.1"]
+	ffff`bb118000	ffff`bb119000	0`00001000	MEM_PRIVATE	MEM_COMMIT	PAGE_EXECUTE_READ	Image	[LC_MONETARY; "/usr/lib/locale/C.utf8/LC_MONETARY"]
+	ffff`bb119000	ffff`bb11b000	0`00002000	MEM_PRIVATE	MEM_COMMIT	PAGE_READWRITE	\<unknown\>	[................]
+	ffff`bb11b000	ffff`bb11c000	0`00001000	MEM_PRIVATE	MEM_COMMIT	PAGE_EXECUTE_READ	Image	[SYS_LC_MESSAGES; "/usr/lib/locale/C.utf8/LC_MESSAGES/SYS_LC_MESSAGES"]
+	ffff`bb11c000	ffff`bb11d000	0`00001000	MEM_PRIVATE	MEM_COMMIT	PAGE_EXECUTE_READ	Image	[LC_PAPER; "/usr/lib/locale/C.utf8/LC_PAPER"]
+	ffff`bb11d000	ffff`bb11e000	0`00001000	MEM_PRIVATE	MEM_COMMIT	PAGE_EXECUTE_READ	Image	[LC_NAME; "/usr/lib/locale/C.utf8/LC_NAME"]
+	ffff`bb11e000	ffff`bb11f000	0`00001000	MEM_PRIVATE	MEM_COMMIT	PAGE_EXECUTE_READ	Image	[LC_ADDRESS; "/usr/lib/locale/C.utf8/LC_ADDRESS"]
+	ffff`bb11f000	ffff`bb120000	0`00001000	MEM_PRIVATE	MEM_COMMIT	PAGE_EXECUTE_READ	Image	[LC_TELEPHONE; "/usr/lib/locale/C.utf8/LC_TELEPHONE"]
+	ffff`bb120000	ffff`bb121000	0`00001000	MEM_PRIVATE	MEM_COMMIT	PAGE_EXECUTE_READ	Image	[LC_MEASUREMENT; "/usr/lib/locale/C.utf8/LC_MEASUREMENT"]
+	ffff`bb121000	ffff`bb122000	0`00001000	MEM_PRIVATE	MEM_COMMIT	PAGE_EXECUTE_READ	Image	[LC_IDENTIFICATION; "/usr/lib/locale/C.utf8/LC_IDENTIFICATION"]
+	ffff`bb122000	ffff`bb124000	0`00002000	MEM_PRIVATE	MEM_COMMIT	PAGE_READWRITE	\<unknown\>	[P...............]
+	ffff`bb124000	ffff`bb126000	0`00002000				\<unknown\>	
+	ffff`bb126000	ffff`bb127000	0`00001000	MEM_PRIVATE	MEM_COMMIT	PAGE_EXECUTE_READ	Image	[linux_vdso_so; "linux-vdso.so.1"]
+	ffff`bb127000	ffff`bb129000	0`00002000	MEM_PRIVATE	MEM_COMMIT	PAGE_READONLY	\<unknown\>	[........z..y..x.]
+	ffff`bb129000	ffff`bb12b000	0`00002000	MEM_PRIVATE	MEM_COMMIT	PAGE_READWRITE	\<unknown\>	[........t.......]
+	ffff`bb12b000	ffff`c5fb8000	0`0ae8d000				\<unknown\>	
+	ffff`c5fb8000	ffff`c5fd9000	0`00021000	MEM_PRIVATE	MEM_COMMIT	PAGE_READWRITE	\<unknown\>	[...............]

VS Code and WSL

- [Get started](#)

- Troubleshooting

  Add to `.bashrc`

  ```
  alias code="'/mnt/c/Users/[USER]/AppData/Local/Programs/Microsoft VS Code/Code.exe'"
  ```

To browse Linux API headers, we recommend using some IDE, for example, Visual Studio Code. It works with WSL but may require some troubleshooting.

Get started
https://learn.microsoft.com/en-us/windows/wsl/tutorials/wsl-vscode

Calling Convention

- **ioctl** (from documentation)

- Actual declaration (`sys/ioctl.h`)

  `extern int ioctl (int __fd, unsigned long int __request, ...) __THROW;`

- `__THROW` is defined in `sys/cdefs.h`

- Argument passing order (simplified)

  - x64 left-to-right via `RDI, RSI, RDX, RCX, R8, R9`, right-to-left `PUSH ...`

  - A64 left-to-right via `X0 - X7, [SP], [SP+8], [SP+16], ...`

- System V Application Binary Interface: AMD64 Architecture Processor Supplement

Except for some API calls that do not need arguments, most API calls require them. How such a source code is implemented on a particular platform is a subject of the so-called calling convention. In the x64 calling convention, the first six parameters are passed via registers from left to right and the rest – via stack (usually implemented via push hence negative **RSP** addressing, it also requires the caller to clear the stack by adjusting **RSP** after the call). In the ARM64 calling convention, the first eight parameters are passed via registers from left to right and the rest – via the stack, too (usually implemented via direct stack addressing in the preallocated region, hence positive **SP** addressing). Both are illustrated on the next two slides, and both platforms are illustrated in Exercise L2. I chose the *ioctl* function as it allows passing many parameters. Usually, Linux API only requires a few parameters. If you are interested in more details, the x64 ABI link has more information about the calling convention used.

ioctl
https://man7.org/linux/man-pages/man2/ioctl.2.html

System V Application Binary Interface: AMD64 Architecture Processor Supplement
https://github.com/hjl-tools/x86-psABI/wiki/x86-64-psABI-1.0.pdf

Parameter Passing (x64)

```
Test8params(int p1, int p2, int p3, int p4, int p5, int p6, int p7, int p8);
```

Caller

```
EDI (p1)
ESI (p2)
EDX (p3)
ESX (p4)
R8D (p5)
R9D (p6)
```

Callee

```
EDI (p1)
ESI (p2)
EDX (p3)
ESX (p4)
R8D (p5)
R9D (p6)
```

call →

```
RSP:      0`p7
RSP+0x8: 0`p8
```

```
RSP:       return address
RSP+0x8:  0`p7
RSP+0x10: 0`p8
```

Parameter Passing (A64)

```
Test10params(int p1, int p2, int p3, int p4, int p5, int p6, int p7, int p8, int p9, int p10);
```

Caller   Callee

W0 (p1)   W0 (p1)
W1 (p2)   W1 (p2)
W2 (p3)   W2 (p3)
W3 (p4)   W3 (p4)
W4 (p5)   W4 (p5)
W5 (p6)   W5 (p6)
W6 (p7)   call   W6 (p7)
W7 (p8)   ───────────────▶   W7 (p8)

SP: 0`p9                                                     SP:    0`p9
SP+8: 0`p10                                                     SP+8: 0`p10

Exercise L2

- **Goal:** Compare calling conventions on x64 and A64 platforms

- **ADDR Patterns:** Call Prologue; Call Parameter; Function Prologue

- \LAPI-Dumps\Exercise-L2-GDB.pdf

- \LAPI-Dumps\Exercise-L2-WinDbg.pdf

Exercise L2 (GDB)

Goal: Compare calling conventions on x64 and A64 platforms.

ADDR Patterns: Call Prologue; Call Parameter; Function Prologue.

1.   Load a core dump *core.8323* and *ioctl-params* executable from the x64 directory:

```
~/LAPI/x64$ gdb -c core.8323 -se ioctl-params
GNU gdb (Debian 8.2.1-2+b3) 8.2.1
Copyright (C) 2018 Free Software Foundation, Inc.
License GPLv3+: GNU GPL version 3 or later <http://gnu.org/licenses/gpl.html>
This is free software: you are free to change and redistribute it.
There is NO WARRANTY, to the extent permitted by law.
Type "show copying" and "show warranty" for details.
This GDB was configured as "x86_64-linux-gnu".
Type "show configuration" for configuration details.
For bug reporting instructions, please see:
<http://www.gnu.org/software/gdb/bugs/>.
Find the GDB manual and other documentation resources online at:
    <http://www.gnu.org/software/gdb/documentation/>.

For help, type "help".
Type "apropos word" to search for commands related to "word"...
Reading symbols from ioctl-params...(no debugging symbols found)...done.
[New LWP 8323]
Core was generated by `./ioctl-params'.
#0  0x00007f77d6368594 in __GI___nanosleep
(requested_time=requested_time@entry=0x7fff5574a3c0, remaining=remaining@entry=0x7fff5574a3c0)
    at ../sysdeps/unix/sysv/linux/nanosleep.c:28
28      ../sysdeps/unix/sysv/linux/nanosleep.c: No such file or directory.
```

2.   Set logging to a file in case of lengthy output from some commands:

```
(gdb) set logging on L2.log
Copying output to L2.log.
```

3.   Disassemble the *main* function and find the call to the *ioctl@plt* function:

```
(gdb) disassemble main
Dump of assembler code for function main:
   0x00005648dd44d145 <+0>:     push    %rbp
   0x00005648dd44d146 <+1>:     mov     %rsp,%rbp
   0x00005648dd44d149 <+4>:     sub     $0x10,%rsp
   0x00005648dd44d14d <+8>:     mov     %edi,-0x4(%rbp)
   0x00005648dd44d150 <+11>:    mov     %rsi,-0x10(%rbp)
   0x00005648dd44d154 <+15>:    mov     $0xffffffff,%edi
   0x00005648dd44d159 <+20>:    callq   0x5648dd44d040 <sleep@plt>
   0x00005648dd44d15e <+25>:    sub     $0x8,%rsp
   0x00005648dd44d162 <+29>:    pushq   $0xa
   0x00005648dd44d164 <+31>:    pushq   $0x9
   0x00005648dd44d166 <+33>:    pushq   $0x8
   0x00005648dd44d168 <+35>:    pushq   $0x7
   0x00005648dd44d16a <+37>:    pushq   $0x6
   0x00005648dd44d16c <+39>:    mov     $0x5,%r9d
   0x00005648dd44d172 <+45>:    mov     $0x4,%r8d
```

```
0x00005648dd44d178 <+51>:    mov    $0x3,%ecx
0x00005648dd44d17d <+56>:    mov    $0x2,%edx
0x00005648dd44d182 <+61>:    mov    $0x1,%esi
0x00005648dd44d187 <+66>:    mov    $0x0,%edi
0x00005648dd44d18c <+71>:    mov    $0x0,%eax
0x00005648dd44d191 <+76>:    callq  0x5648dd44d030 <ioctl@plt>
0x00005648dd44d196 <+81>:    add    $0x30,%rsp
0x00005648dd44d19a <+85>:    leaveq
0x00005648dd44d19b <+86>:    retq
End of assembler dump.
```

Note: We see the first 6 parameters are passed via registers, then the next 5 parameters are pushed to stack. The caller has to adjust the stack pointer (RSP) by 0x30 (6*8) instead of (5*8) because there was an additional *sub 8* instruction.

4.   Load a core dump *core.45713* and *ioctl-params* executable from the A64 directory:

```
~/LAPI/x64$ cd ../A64
```

```
~/LAPI/A64$ gdb-multiarch -c core.45713 -se ioctl-params
GNU gdb (Debian 8.2.1-2+b3) 8.2.1
Copyright (C) 2018 Free Software Foundation, Inc.
License GPLv3+: GNU GPL version 3 or later <http://gnu.org/licenses/gpl.html>
This is free software: you are free to change and redistribute it.
There is NO WARRANTY, to the extent permitted by law.
Type "show copying" and "show warranty" for details.
This GDB was configured as "x86_64-linux-gnu".
Type "show configuration" for configuration details.
For bug reporting instructions, please see:
<http://www.gnu.org/software/gdb/bugs/>.
Find the GDB manual and other documentation resources online at:
    <http://www.gnu.org/software/gdb/documentation/>.

For help, type "help".
Type "apropos word" to search for commands related to "word"...
Reading symbols from ioctl-params...(no debugging symbols found)...done.
[New LWP 45713]

warning: Could not load shared library symbols for 2 libraries, e.g.
/lib/aarch64-linux-gnu/libc.so.6.
Use the "info sharedlibrary" command to see the complete listing.
Do you need "set solib-search-path" or "set sysroot"?
Core was generated by `./ioctl-params'.
#0  0x0000ffff832b189c in ?? ()
```

```
(gdb) set solib-search-path .
Reading symbols from /home/coredump/LAPI/A64/libc.so.6...(no debugging symbols found)...done.
Reading symbols from /home/coredump/LAPI/A64/ld-linux-aarch64.so.1...(no debugging symbols found)...done.
```

5.   Set logging to a file in case of lengthy output from some commands:

```
(gdb) set logging on L2.log
Copying output to L2.log.
```

6. Disassemble the *main* function and find the branch and link to the *ioctl@plt* function:

```
(gdb) disassemble main
Dump of assembler code for function main:
   0x0000aaaab63007d4 <+0>:     sub     sp, sp, #0x40
   0x0000aaaab63007d8 <+4>:     stp     x29, x30, [sp, #32]
   0x0000aaaab63007dc <+8>:     add     x29, sp, #0x20
   0x0000aaaab63007e0 <+12>:    str     w0, [sp, #60]
   0x0000aaaab63007e4 <+16>:    str     x1, [sp, #48]
   0x0000aaaab63007e8 <+20>:    mov     w0, #0xffffffff              // #-1
   0x0000aaaab63007ec <+24>:    bl      0xaaaab6300650 <sleep@plt>
   0x0000aaaab63007f0 <+28>:    mov     w0, #0xa                     // #10
   0x0000aaaab63007f4 <+32>:    str     w0, [sp, #16]
   0x0000aaaab63007f8 <+36>:    mov     w0, #0x9                     // #9
   0x0000aaaab63007fc <+40>:    str     w0, [sp, #8]
   0x0000aaaab6300800 <+44>:    mov     w0, #0x8                     // #8
   0x0000aaaab6300804 <+48>:    str     w0, [sp]
   0x0000aaaab6300808 <+52>:    mov     w7, #0x7                     // #7
   0x0000aaaab630080c <+56>:    mov     w6, #0x6                     // #6
   0x0000aaaab6300810 <+60>:    mov     w5, #0x5                     // #5
   0x0000aaaab6300814 <+64>:    mov     w4, #0x4                     // #4
   0x0000aaaab6300818 <+68>:    mov     w3, #0x3                     // #3
   0x0000aaaab630081c <+72>:    mov     w2, #0x2                     // #2
   0x0000aaaab6300820 <+76>:    mov     x1, #0x1                     // #1
   0x0000aaaab6300824 <+80>:    mov     w0, #0x0                     // #0
   0x0000aaaab6300828 <+84>:    bl      0xaaaab6300680 <ioctl@plt>
   0x0000aaaab630082c <+88>:    ldp     x29, x30, [sp, #32]
   0x0000aaaab6300830 <+92>:    add     sp, sp, #0x40
   0x0000aaaab6300834 <+96>:    ret
End of assembler dump.
```

Note: We see the first 8 parameters are passed via registers, then the next 3 parameters are saved in the already preallocated stack, so the caller doesn't have to adjust the stack pointer (SP) right after the call.

Exercise L2 (WinDbg)

Goal: Compare calling conventions on x64 and A64 platforms.

ADDR Patterns: Call Prologue; Call Parameter; Function Prologue.

1.   Launch WinDbg and load a core dump *core.45713* from the A64 directory:

```
Loading Dump File [C:\LAPI\A64\core.45713]
64-bit machine not using 64-bit API

************* Path validation summary **************
Response                      Time (ms)      Location
Deferred                                     srv*
Symbol search path is: srv*
Executable search path is:
Generic Unix Version 0 UP Free ARM 64-bit (AArch64)
System Uptime: not available
Process Uptime: not available
....
*** WARNING: Unable to verify timestamp for libc.so.6
libc_so+0xb189c:
0000ffff`832b189c d4000001 svc          #0
```

2.   Set the symbol path and logging to a file in case of lengthy output from some commands:

```
0:000> .sympath+ C:\LAPI\A64
Symbol search path is: srv*;C:\LAPI\A64
Expanded Symbol search path is: cache*;SRV*https://msdl.microsoft.com/download/symbols;c:\
lapi\a64

************* Path validation summary **************
Response                      Time (ms)      Location
Deferred                                     srv*
OK                                           C:\LAPI\A64
*** WARNING: Unable to verify timestamp for libc.so.6

0:000> .reload
...*** WARNING: Unable to verify timestamp for libc.so.6
.

************* Symbol Loading Error Summary **************
Module name          Error
libc.so              The system cannot find the file specified

You can troubleshoot most symbol related issues by turning on symbol loading diagnostics (!sym
noisy) and repeating the command that caused symbols to be loaded.
You should also verify that your symbol search path (.sympath) is correct.

0:000> .logopen C:\LAPI\A64\L2-WinDbg.log
Opened log file 'C:\LAPI\A64\L2-WinDbg.log'
```

3.   Disassemble the *main* function and find the branch and link to the address 0xaaaab6300680, identified as *ioctl@plt* function in the GDB exercise:

```
0:000> uf main
ioctl_params!main:
0000aaaa`b63007d4 d10103ff sub     sp,sp,#0x40
0000aaaa`b63007d8 a9027bfd stp     fp,lr,[sp,#0x20]
0000aaaa`b63007dc 910083fd add     fp,sp,#0x20
0000aaaa`b63007e0 b9003fe0 str     w0,[sp,#0x3C]
0000aaaa`b63007e4 f9001be1 str     x1,[sp,#0x30]
0000aaaa`b63007e8 12800000 mov     w0,#-1
0000aaaa`b63007ec 97ffff99 bl      ioctl_params+0x650 (0000aaaa`b6300650)
0000aaaa`b63007f0 52800140 mov     w0,#0xA
0000aaaa`b63007f4 b90013e0 str     w0,[sp,#0x10]
0000aaaa`b63007f8 52800120 mov     w0,#9
0000aaaa`b63007fc b9000be0 str     w0,[sp,#8]
0000aaaa`b6300800 52800100 mov     w0,#8
0000aaaa`b6300804 b90003e0 str     w0,[sp]
0000aaaa`b6300808 528000e7 mov     w7,#7
0000aaaa`b630080c 528000c6 mov     w6,#6
0000aaaa`b6300810 528000a5 mov     w5,#5
0000aaaa`b6300814 52800084 mov     w4,#4
0000aaaa`b6300818 52800063 mov     w3,#3
0000aaaa`b630081c 52800042 mov     w2,#2
0000aaaa`b6300820 d2800021 mov     x1,#1
0000aaaa`b6300824 52800000 mov     w0,#0
0000aaaa`b6300828 97ffff96 bl      ioctl_params+0x680 (0000aaaa`b6300680)
0000aaaa`b630082c a9427bfd ldp     fp,lr,[sp,#0x20]
0000aaaa`b6300830 910103ff add     sp,sp,#0x40
0000aaaa`b6300834 d65f03c0 ret
```

Note: We see the first 8 parameters are passed via registers, then the next 3 parameters are saved in the already preallocated stack, so the caller doesn't have to adjust the stack pointer (SP) right after the call.

Static Linkage

- ⦿ Example

  - core.18256 from /LAPI/x64/static-example

    ```
    main:
    ...
    callq  0x43c1f0 <sleep>
    ...
    callq  0x43d6e0 <ioctl>
    ...
    ```

For comparison, I created a statically linked example to show that in such a case, Linux API is called directly.

Shared Libraries

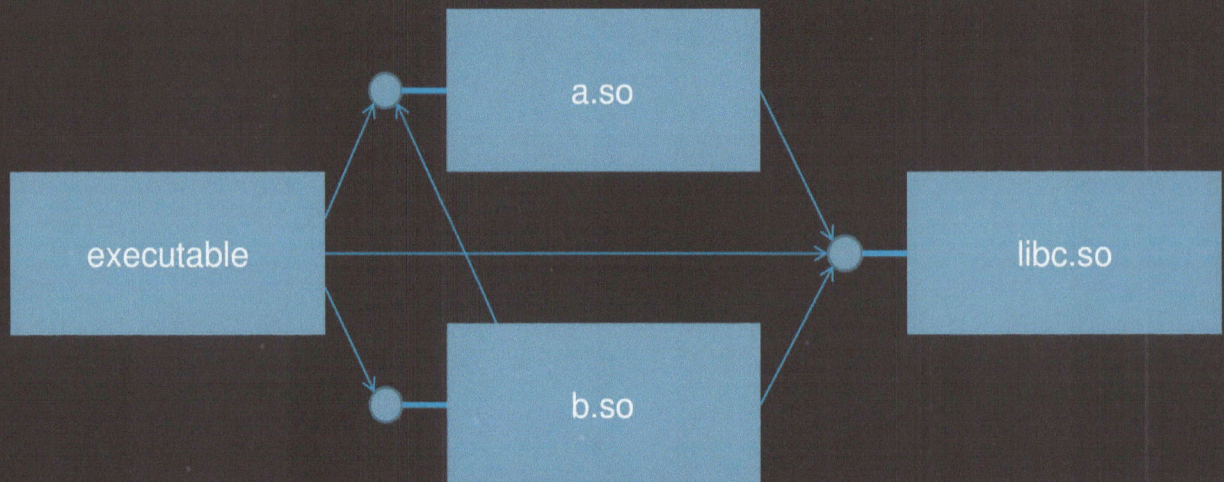

GDB Commands

```
(gdb) info sharedlibrary

(gdb) info files

(gdb) info functions pattern
```

WinDbg Commands

```
0:000> lm

0:000> x mpattern!pattern

0:000> x *!pattern
```

© 2023 Software Diagnostics Services

Shared libraries can import functions from other shared libraries. But function names are not unique. Several shared libraries may export the same function names. Such functions may or may not have the same parameters. So, function syntax, semantics, and pragmatics are subject to API design. Also, not all functions from shared libraries are imported – only those that are actually used in code.

Modules and Analysis Patterns

- ◉ Module memory analysis patterns

  - Module Collection
  - Coupled Modules
  - Duplicated Module

- ◉ Namespace malware analysis pattern

There are many memory analysis patterns related to modules (executables and shared libraries). On this slide, I list a few of them related to module dependencies and imported functions. **Module Collection** lists loaded modules. **Coupled Modules** is about module dependencies. Sometimes, the same module name may be loaded twice, for example, from different locations, the so-called **Duplicated Module**. Imported functions constitute the so-called **Namespace** malware analysis pattern which can give hints at overall module functionality and purpose. For example, an innocuous module name may have functions imported from LIBC that reveal potential network connectivity.

Note: all referenced patterns from dumpanalysis.org (and hundreds of others) are also available in **Memory Dump Analysis Anthology** volumes or **Encyclopedia of Crash Dump Analysis Patterns** (see References slides).

Exercise L3

- **Goal:** Explore shared libraries and their dependencies

- **Memory Analysis Patterns:** Module Collection; Coupled Modules; Value References

- **Malware Analysis Patterns:** Namespace

- \LAPI-Dumps\Exercise-L3-GDB.pdf

- \LAPI-Dumps\Exercise-L3-WinDbg.pdf

Exercise L3 (GDB)

Goal: Explore shared libraries and their dependencies.

Memory Analysis Patterns: Module Collection; Coupled Modules.

Malware Analysis Patterns: Namespace.

1. Check library dependencies of *bash* using *ldd*:

```
~/LAPI/x64$ ldd -v bash
        linux-vdso.so.1 (0x00007ffd5cd53000)
        libtinfo.so.6 => /lib/x86_64-linux-gnu/libtinfo.so.6 (0x00007fb8f43a6000)
        libdl.so.2 => /lib/x86_64-linux-gnu/libdl.so.2 (0x00007fb8f43a1000)
        libc.so.6 => /lib/x86_64-linux-gnu/libc.so.6 (0x00007fb8f41e1000)
        /lib64/ld-linux-x86-64.so.2 (0x00007fb8f4509000)

        Version information:
        ./bash:
                libdl.so.2 (GLIBC_2.2.5) => /lib/x86_64-linux-gnu/libdl.so.2
                libtinfo.so.6 (NCURSES6_TINFO_5.0.19991023) => /lib/x86_64-linux-gnu/libtinfo.so.6
                libc.so.6 (GLIBC_2.11) => /lib/x86_64-linux-gnu/libc.so.6
                libc.so.6 (GLIBC_2.14) => /lib/x86_64-linux-gnu/libc.so.6
                libc.so.6 (GLIBC_2.8) => /lib/x86_64-linux-gnu/libc.so.6
                libc.so.6 (GLIBC_2.15) => /lib/x86_64-linux-gnu/libc.so.6
                libc.so.6 (GLIBC_2.4) => /lib/x86_64-linux-gnu/libc.so.6
                libc.so.6 (GLIBC_2.3.4) => /lib/x86_64-linux-gnu/libc.so.6
                libc.so.6 (GLIBC_2.3) => /lib/x86_64-linux-gnu/libc.so.6
                libc.so.6 (GLIBC_2.2.5) => /lib/x86_64-linux-gnu/libc.so.6
        /lib/x86_64-linux-gnu/libtinfo.so.6:
                libc.so.6 (GLIBC_2.3) => /lib/x86_64-linux-gnu/libc.so.6
                libc.so.6 (GLIBC_2.14) => /lib/x86_64-linux-gnu/libc.so.6
                libc.so.6 (GLIBC_2.16) => /lib/x86_64-linux-gnu/libc.so.6
                libc.so.6 (GLIBC_2.4) => /lib/x86_64-linux-gnu/libc.so.6
                libc.so.6 (GLIBC_2.3.4) => /lib/x86_64-linux-gnu/libc.so.6
                libc.so.6 (GLIBC_2.2.5) => /lib/x86_64-linux-gnu/libc.so.6
        /lib/x86_64-linux-gnu/libdl.so.2:
                ld-linux-x86-64.so.2 (GLIBC_PRIVATE) => /lib64/ld-linux-x86-64.so.2
                libc.so.6 (GLIBC_PRIVATE) => /lib/x86_64-linux-gnu/libc.so.6
                libc.so.6 (GLIBC_2.4) => /lib/x86_64-linux-gnu/libc.so.6
                libc.so.6 (GLIBC_2.2.5) => /lib/x86_64-linux-gnu/libc.so.6
        /lib/x86_64-linux-gnu/libc.so.6:
                ld-linux-x86-64.so.2 (GLIBC_2.3) => /lib64/ld-linux-x86-64.so.2
                ld-linux-x86-64.so.2 (GLIBC_PRIVATE) => /lib64/ld-linux-x86-64.so.2
```

2. Load a core dump *core.9* and *bash* executable from the x64 directory:

```
~/LAPI/x64$ gdb -c core.9 -se bash
GNU gdb (Debian 8.2.1-2+b3) 8.2.1
Copyright (C) 2018 Free Software Foundation, Inc.
License GPLv3+: GNU GPL version 3 or later <http://gnu.org/licenses/gpl.html>
This is free software: you are free to change and redistribute it.
There is NO WARRANTY, to the extent permitted by law.
Type "show copying" and "show warranty" for details.
This GDB was configured as "x86_64-linux-gnu".
Type "show configuration" for configuration details.
For bug reporting instructions, please see:
<http://www.gnu.org/software/gdb/bugs/>.
Find the GDB manual and other documentation resources online at:
```

```
        <http://www.gnu.org/software/gdb/documentation/>.

For help, type "help".
Type "apropos word" to search for commands related to "word"...
Reading symbols from bash...(no debugging symbols found)...done.

warning: core file may not match specified executable file.
[New LWP 9]
Core was generated by `-bash'.
#0  0x00007f3e9f7492d7 in __GI__waitpid (pid=-1, stat_loc=0x7ffcbd661ad0, options=10) at
../sysdeps/unix/sysv/linux/waitpid.c:30
30        ../sysdeps/unix/sysv/linux/waitpid.c: No such file or directory.
```

3.   Set logging to a file in case of lengthy output from some commands:

```
(gdb) set logging on L3.log
Copying output to L3.log.
```

4.   Find the shared library the _dl_update_slotinfo function is from:

```
(gdb) info address _dl_update_slotinfo
Symbol "_dl_update_slotinfo" is a function at address 0x7f3e9f895100.
```

```
(gdb) info sharedlibrary
From                To                  Syms Read   Shared Object Library
0x00007f3e9f856950  0x00007f3e9f863dc8  Yes (*)     /lib/x86_64-linux-gnu/libtinfo.so.6
0x00007f3e9f844130  0x00007f3e9f844eb5  Yes         /lib/x86_64-linux-gnu/libdl.so.2
0x00007f3e9f6a5320  0x00007f3e9f7eb14b  Yes         /lib/x86_64-linux-gnu/libc.so.6
0x00007f3e9f884090  0x00007f3e9f8a1b50  Yes         /lib64/ld-linux-x86-64.so.2
0x00007f3e9f37e300  0x00007f3e9f384578  Yes         /lib/x86_64-linux-gnu/libnss_files.so.2
(*): Shared library is missing debugging information.
```

5.   Disassemble the _dl_update_slotinfo function and find the call to the free function:

```
(gdb) disassemble _dl_update_slotinfo
Dump of assembler code for function _dl_update_slotinfo:
   0x00007f3e9f895100 <+0>:    push   %r15
   0x00007f3e9f895102 <+2>:    push   %r14
   0x00007f3e9f895104 <+4>:    push   %r13
   0x00007f3e9f895106 <+6>:    push   %r12
   0x00007f3e9f895108 <+8>:    push   %rbp
   0x00007f3e9f895109 <+9>:    push   %rbx
   0x00007f3e9f89510a <+10>:   sub    $0x38,%rsp
   0x00007f3e9f89510e <+14>:   mov    %fs:0x8,%r10
   0x00007f3e9f895117 <+23>:   mov    0x16e8a(%rip),%r14      # 0x7f3e9f8abfa8
<_rtld_global+3912>
   0x00007f3e9f89511e <+30>:   mov    %r10,%r15
   0x00007f3e9f895121 <+33>:   mov    %rdi,%rdx
   0x00007f3e9f895124 <+36>:   mov    (%r14),%rcx
   0x00007f3e9f895127 <+39>:   mov    %r14,%rsi
   0x00007f3e9f89512a <+42>:   cmp    %rcx,%rdi
   0x00007f3e9f89512d <+45>:   jb     0x7f3e9f895147 <_dl_update_slotinfo+71>
   0x00007f3e9f89512f <+47>:   mov    %rcx,%r9
   0x00007f3e9f895132 <+50>:   nopw   0x0(%rax,%rax,1)
   0x00007f3e9f895138 <+56>:   mov    0x8(%rsi),%rsi
   0x00007f3e9f89513c <+60>:   sub    %r9,%rdx
   0x00007f3e9f89513f <+63>:   mov    (%rsi),%r9
   0x00007f3e9f895142 <+66>:   cmp    %rdx,%r9
   0x00007f3e9f895145 <+69>:   jbe    0x7f3e9f895138 <_dl_update_slotinfo+56>
```

```
0x00007f3e9f895147 <+71>:    add     $0x1,%rdx
0x00007f3e9f89514b <+75>:    shl     $0x4,%rdx
0x00007f3e9f89514f <+79>:    mov     (%rsi,%rdx,1),%rbx
0x00007f3e9f895153 <+83>:    cmp     %rbx,(%r10)
0x00007f3e9f895156 <+86>:    jae     0x7f3e9f8952bd <_dl_update_slotinfo+445>
0x00007f3e9f89515c <+92>:    mov     %rdi,0x20(%rsp)
0x00007f3e9f895161 <+97>:    movq    $0x0,0x18(%rsp)
0x00007f3e9f89516a <+106>:   movq    $0x0,0x8(%rsp)
0x00007f3e9f895173 <+115>:   nopl    0x0(%rax,%rax,1)
0x00007f3e9f895178 <+120>:   mov     0x18(%rsp),%rax
0x00007f3e9f89517d <+125>:   xor     %r13d,%r13d
0x00007f3e9f895180 <+128>:   test    %rax,%rax
0x00007f3e9f895183 <+131>:   sete    %r13b
0x00007f3e9f895187 <+135>:   cmp     %rcx,%r13
0x00007f3e9f89518a <+138>:   jae     0x7f3e9f895290 <_dl_update_slotinfo+400>
0x00007f3e9f895190 <+144>:   shl     $0x4,%rax
0x00007f3e9f895194 <+148>:   mov     %rax,0x28(%rsp)
0x00007f3e9f895199 <+153>:   jmpq    0x7f3e9f895226 <_dl_update_slotinfo+294>
0x00007f3e9f89519e <+158>:   xchg    %ax,%ax
0x00007f3e9f8951a0 <+160>:   mov     0x18(%rsp),%rax
0x00007f3e9f8951a5 <+165>:   mov     0x450(%r12),%rbp
0x00007f3e9f8951ad <+173>:   lea     (%rax,%r13,1),%rcx
--Type <RET> for more, q to quit, c to continue without paging--
0x00007f3e9f8951b1 <+177>:   cmp     %rcx,%rbp
0x00007f3e9f8951b4 <+180>:   jne     0x7f3e9f8952c8 <_dl_update_slotinfo+456>
0x00007f3e9f8951ba <+186>:   cmp     %rbp,-0x10(%r15)
0x00007f3e9f8951be <+190>:   jae     0x7f3e9f8951de <_dl_update_slotinfo+222>
0x00007f3e9f8951c0 <+192>:   mov     %r15,%rdi
0x00007f3e9f8951c3 <+195>:   callq   0x7f3e9f894730 <_dl_resize_dtv>
0x00007f3e9f8951c8 <+200>:   mov     %rax,%r15
0x00007f3e9f8951cb <+203>:   cmp     %rbp,-0x10(%rax)
0x00007f3e9f8951cf <+207>:   jb      0x7f3e9f8952e7 <_dl_update_slotinfo+487>
0x00007f3e9f8951d5 <+213>:   mov     %rax,%fs:0x8
0x00007f3e9f8951de <+222>:   mov     %rbp,%rcx
0x00007f3e9f8951e1 <+225>:   shl     $0x4,%rcx
0x00007f3e9f8951e5 <+229>:   add     %r15,%rcx
0x00007f3e9f8951e8 <+232>:   mov     0x8(%rcx),%rdi
0x00007f3e9f8951ec <+236>:   mov     %rcx,0x10(%rsp)
0x00007f3e9f8951f1 <+241>:   callq   0x7f3e9f884080 <free@plt>
0x00007f3e9f8951f6 <+246>:   mov     0x10(%rsp),%rcx
0x00007f3e9f8951fb <+251>:   cmp     %rbp,0x20(%rsp)
0x00007f3e9f895200 <+256>:   cmovne  0x8(%rsp),%r12
0x00007f3e9f895206 <+262>:   movq    $0xffffffffffffffff,(%rcx)
0x00007f3e9f89520d <+269>:   movq    $0x0,0x8(%rcx)
0x00007f3e9f895215 <+277>:   mov     %r12,0x8(%rsp)
0x00007f3e9f89521a <+282>:   mov     (%r14),%rcx
0x00007f3e9f89521d <+285>:   add     $0x1,%r13
0x00007f3e9f895221 <+289>:   cmp     %r13,%rcx
0x00007f3e9f895224 <+292>:   jbe     0x7f3e9f895290 <_dl_update_slotinfo+400>
0x00007f3e9f895226 <+294>:   mov     %r13,%rsi
0x00007f3e9f895229 <+297>:   shl     $0x4,%rsi
0x00007f3e9f89522d <+301>:   mov     0x10(%r14,%rsi,1),%rcx
0x00007f3e9f895232 <+306>:   cmp     %rcx,%rbx
0x00007f3e9f895235 <+309>:   jb      0x7f3e9f89521a <_dl_update_slotinfo+282>
0x00007f3e9f895237 <+311>:   cmp     %rcx,(%r15)
0x00007f3e9f89523a <+314>:   jae     0x7f3e9f89521a <_dl_update_slotinfo+282>
0x00007f3e9f89523c <+316>:   mov     0x18(%r14,%rsi,1),%r12
0x00007f3e9f895241 <+321>:   test    %r12,%r12
0x00007f3e9f895244 <+324>:   jne     0x7f3e9f8951a0 <_dl_update_slotinfo+160>
0x00007f3e9f89524a <+330>:   mov     0x18(%rsp),%rax
```

```
0x00007f3e9f89524f <+335>:    lea     (%rax,%r13,1),%rcx
0x00007f3e9f895253 <+339>:    cmp     %rcx,-0x10(%r15)
0x00007f3e9f895257 <+343>:    jb      0x7f3e9f89521a <_dl_update_slotinfo+282>
0x00007f3e9f895259 <+345>:    add     0x28(%rsp),%rsi
0x00007f3e9f89525e <+350>:    add     $0x1,%r13
0x00007f3e9f895262 <+354>:    lea     (%r15,%rsi,1),%rbp
0x00007f3e9f895266 <+358>:    mov     0x8(%rbp),%rdi
0x00007f3e9f89526a <+362>:    callq   0x7f3e9f884080 <free@plt>
--Type <RET> for more, q to quit, c to continue without paging--
0x00007f3e9f89526f <+367>:    mov     (%r14),%rcx
0x00007f3e9f895272 <+370>:    movq    $0xffffffffffffffff,0x0(%rbp)
0x00007f3e9f89527a <+378>:    movq    $0x0,0x8(%rbp)
0x00007f3e9f895282 <+386>:    cmp     %r13,%rcx
0x00007f3e9f895285 <+389>:    ja      0x7f3e9f895226 <_dl_update_slotinfo+294>
0x00007f3e9f895287 <+391>:    nopw    0x0(%rax,%rax,1)
0x00007f3e9f895290 <+400>:    mov     0x8(%r14),%r14
0x00007f3e9f895294 <+404>:    add     %rcx,0x18(%rsp)
0x00007f3e9f895299 <+409>:    test    %r14,%r14
0x00007f3e9f89529c <+412>:    je      0x7f3e9f8952a6 <_dl_update_slotinfo+422>
0x00007f3e9f89529e <+414>:    mov     (%r14),%rcx
0x00007f3e9f8952a1 <+417>:    jmpq    0x7f3e9f895178 <_dl_update_slotinfo+120>
0x00007f3e9f8952a6 <+422>:    mov     %rbx,(%r15)
0x00007f3e9f8952a9 <+425>:    mov     0x8(%rsp),%rax
0x00007f3e9f8952ae <+430>:    add     $0x38,%rsp
0x00007f3e9f8952b2 <+434>:    pop     %rbx
0x00007f3e9f8952b3 <+435>:    pop     %rbp
0x00007f3e9f8952b4 <+436>:    pop     %r12
0x00007f3e9f8952b6 <+438>:    pop     %r13
0x00007f3e9f8952b8 <+440>:    pop     %r14
0x00007f3e9f8952ba <+442>:    pop     %r15
0x00007f3e9f8952bc <+444>:    retq
0x00007f3e9f8952bd <+445>:    movq    $0x0,0x8(%rsp)
0x00007f3e9f8952c6 <+454>:    jmp     0x7f3e9f8952a9 <_dl_update_slotinfo+425>
0x00007f3e9f8952c8 <+456>:    lea     0x111f1(%rip),%rcx        # 0x7f3e9f8a64c0
<__PRETTY_FUNCTION__.9851>
0x00007f3e9f8952cf <+463>:    mov     $0x2b9,%edx
0x00007f3e9f8952d4 <+468>:    lea     0xe43e(%rip),%rsi         # 0x7f3e9f8a3719
0x00007f3e9f8952db <+475>:    lea     0xe455(%rip),%rdi         # 0x7f3e9f8a3737
0x00007f3e9f8952e2 <+482>:    callq   0x7f3e9f89c090 <__GI___assert_fail>
0x00007f3e9f8952e7 <+487>:    lea     0x111d2(%rip),%rcx        # 0x7f3e9f8a64c0
<__PRETTY_FUNCTION__.9851>
0x00007f3e9f8952ee <+494>:    mov     $0x2bf,%edx
0x00007f3e9f8952f3 <+499>:    lea     0xe41f(%rip),%rsi         # 0x7f3e9f8a3719
0x00007f3e9f8952fa <+506>:    lea     0xe44b(%rip),%rdi         # 0x7f3e9f8a374c
0x00007f3e9f895301 <+513>:    callq   0x7f3e9f89c090 <__GI___assert_fail>
End of assembler dump.
```

6.   Find the section where *free@plt* is located, the real *free* address is stored, and its shared library:

```
(gdb) disassemble 0x7f3e9f884080
Dump of assembler code for function free@plt:
   0x00007f3e9f884080 <+0>:    jmpq    *0x26f6a(%rip)        # 0x7f3e9f8aaff0
   0x00007f3e9f884086 <+6>:    xchg    %ax,%ax
End of assembler dump.

(gdb) x/a 0x7f3e9f8aaff0
0x7f3e9f8aaff0:  0x7f3e9f707b60 <__GI___libc_free>
```

```
(gdb) info files
Symbols from "/home/coredump/LAPI/x64/bash".
Local core dump file:
        `/home/coredump/LAPI/x64/core.9', file type elf64-x86-64.
        0x000055e16515d000 - 0x000055e165160000 is load1
        0x000055e165160000 - 0x000055e165169000 is load2
        0x000055e165169000 - 0x000055e165173000 is load3
        0x000055e1667f5000 - 0x000055e166879000 is load4
        0x00007f3e9f388000 - 0x00007f3e9f389000 is load5
        0x00007f3e9f389000 - 0x00007f3e9f38a000 is load6
        0x00007f3e9f38a000 - 0x00007f3e9f390000 is load7
        0x00007f3e9f680000 - 0x00007f3e9f683000 is load8
        0x00007f3e9f839000 - 0x00007f3e9f83d000 is load9
        0x00007f3e9f83d000 - 0x00007f3e9f83f000 is load10
        0x00007f3e9f83f000 - 0x00007f3e9f843000 is load11
        0x00007f3e9f846000 - 0x00007f3e9f847000 is load12
        0x00007f3e9f847000 - 0x00007f3e9f848000 is load13
        0x00007f3e9f871000 - 0x00007f3e9f875000 is load14
        0x00007f3e9f875000 - 0x00007f3e9f876000 is load15
        0x00007f3e9f876000 - 0x00007f3e9f878000 is load16
        0x00007f3e9f8aa000 - 0x00007f3e9f8ab000 is load17
        0x00007f3e9f8ab000 - 0x00007f3e9f8ac000 is load18
        0x00007f3e9f8ac000 - 0x00007f3e9f8ad000 is load19
        0x00007ffcbd642000 - 0x00007ffcbd663000 is load20
        0x00007ffcbd7ea000 - 0x00007ffcbd7eb000 is load21
Local exec file:
        `/home/coredump/LAPI/x64/bash', file type elf64-x86-64.
        Entry point: 0x55e16507a630
        0x000055e16504b2a8 - 0x000055e16504b2c4 is .interp
        0x000055e16504b2c4 - 0x000055e16504b2e4 is .note.ABI-tag
        0x000055e16504b2e4 - 0x000055e16504b308 is .note.gnu.build-id
        0x000055e16504b308 - 0x000055e16504fdb0 is .gnu.hash
        0x000055e16504fdb0 - 0x000055e16505e1b0 is .dynsym
        0x000055e16505e1b0 - 0x000055e1650678d8 is .dynstr
        0x000055e1650678d8 - 0x000055e165068bd8 is .gnu.version
        0x000055e165068bd8 - 0x000055e165068ca8 is .gnu.version_r
        0x000055e165068ca8 - 0x000055e165076928 is .rela.dyn
        0x000055e165076928 - 0x000055e165077d98 is .rela.plt
        0x000055e165078000 - 0x000055e165078017 is .init
        0x000055e165078020 - 0x000055e165078dd0 is .plt
        0x000055e165078dd0 - 0x000055e165078de8 is .plt.got
        0x000055e165078df0 - 0x000055e165125781 is .text
        0x000055e165125784 - 0x000055e16512578d is .fini
        0x000055e165126000 - 0x000055e16513f930 is .rodata
        0x000055e16513f930 - 0x000055e165143df4 is .eh_frame_hdr
        0x000055e165143df8 - 0x000055e16515b730 is .eh_frame
        0x000055e16515d3f0 - 0x000055e16515d3f8 is .init_array
        0x000055e16515d3f8 - 0x000055e16515d400 is .fini_array
        0x000055e16515d400 - 0x000055e16515fcf0 is .data.rel.ro
--Type <RET> for more, q to quit, c to continue without paging--
        0x000055e16515fcf0 - 0x000055e16515fef0 is .dynamic
        0x000055e16515fef0 - 0x000055e16515fff0 is .got
        0x000055e165160000 - 0x000055e1651606e8 is .got.plt
        0x000055e165160700 - 0x000055e165168d04 is .data
        0x000055e165168d20 - 0x000055e165172998 is .bss
        0x00007f3e9f848238 - 0x00007f3e9f84825c is .note.gnu.build-id in /lib/x86_64-linux-gnu/libtinfo.so.6
        0x00007f3e9f848260 - 0x00007f3e9f848a78 is .gnu.hash in /lib/x86_64-linux-gnu/libtinfo.so.6
        0x00007f3e9f848a78 - 0x00007f3e9f84a7a0 is .dynsym in /lib/x86_64-linux-gnu/libtinfo.so.6
        0x00007f3e9f84a7a0 - 0x00007f3e9f84b71a is .dynstr in /lib/x86_64-linux-gnu/libtinfo.so.6
        0x00007f3e9f84b71a - 0x00007f3e9f84b988 is .gnu.version in /lib/x86_64-linux-gnu/libtinfo.so.6
        0x00007f3e9f84b988 - 0x00007f3e9f84bd34 is .gnu.version_d in /lib/x86_64-linux-gnu/libtinfo.so.6
        0x00007f3e9f84bd38 - 0x00007f3e9f84bda8 is .gnu.version_r in /lib/x86_64-linux-gnu/libtinfo.so.6
        0x00007f3e9f84bda8 - 0x00007f3e9f854d48 is .rela.dyn in /lib/x86_64-linux-gnu/libtinfo.so.6
        0x00007f3e9f854d48 - 0x00007f3e9f855ae0 is .rela.plt in /lib/x86_64-linux-gnu/libtinfo.so.6
        0x00007f3e9f856000 - 0x00007f3e9f856017 is .init in /lib/x86_64-linux-gnu/libtinfo.so.6
        0x00007f3e9f856020 - 0x00007f3e9f856940 is .plt in /lib/x86_64-linux-gnu/libtinfo.so.6
        0x00007f3e9f856940 - 0x00007f3e9f856950 is .plt.got in /lib/x86_64-linux-gnu/libtinfo.so.6
        0x00007f3e9f856950 - 0x00007f3e9f863dc8 is .text in /lib/x86_64-linux-gnu/libtinfo.so.6
        0x00007f3e9f863dc8 - 0x00007f3e9f863dd1 is .fini in /lib/x86_64-linux-gnu/libtinfo.so.6
        0x00007f3e9f864000 - 0x00007f3e9f86d9a5 is .rodata in /lib/x86_64-linux-gnu/libtinfo.so.6
        0x00007f3e9f86d9a8 - 0x00007f3e9f86e13c is .eh_frame_hdr in /lib/x86_64-linux-gnu/libtinfo.so.6
        0x00007f3e9f86e140 - 0x00007f3e9f8706b0 is .eh_frame in /lib/x86_64-linux-gnu/libtinfo.so.6
        0x00007f3e9f8718d0 - 0x00007f3e9f8718d8 is .init_array in /lib/x86_64-linux-gnu/libtinfo.so.6
        0x00007f3e9f8718d8 - 0x00007f3e9f8718e0 is .fini_array in /lib/x86_64-linux-gnu/libtinfo.so.6
        0x00007f3e9f8718e0 - 0x00007f3e9f874848 is .data.rel.ro in /lib/x86_64-linux-gnu/libtinfo.so.6
```

```
        0x00007f3e9f874848 - 0x00007f3e9f874a58 is .dynamic in /lib/x86_64-linux-gnu/libtinfo.so.6
        0x00007f3e9f874a58 - 0x00007f3e9f874ff8 is .got in /lib/x86_64-linux-gnu/libtinfo.so.6
        0x00007f3e9f875000 - 0x00007f3e9f8754f4 is .data in /lib/x86_64-linux-gnu/libtinfo.so.6
        0x00007f3e9f875500 - 0x00007f3e9f875980 is .bss in /lib/x86_64-linux-gnu/libtinfo.so.6
        0x00007f3e9f843238 - 0x00007f3e9f84325c is .note.gnu.build-id in /lib/x86_64-linux-gnu/libdl.so.2
        0x00007f3e9f84325c - 0x00007f3e9f84327c is .note.ABI-tag in /lib/x86_64-linux-gnu/libdl.so.2
        0x00007f3e9f843280 - 0x00007f3e9f843348 is .gnu.hash in /lib/x86_64-linux-gnu/libdl.so.2
        0x00007f3e9f843348 - 0x00007f3e9f843750 is .dynsym in /lib/x86_64-linux-gnu/libdl.so.2
        0x00007f3e9f843750 - 0x00007f3e9f843989 is .dynstr in /lib/x86_64-linux-gnu/libdl.so.2
        0x00007f3e9f84398a - 0x00007f3e9f8439e0 is .gnu.version in /lib/x86_64-linux-gnu/libdl.so.2
        0x00007f3e9f8439e0 - 0x00007f3e9f843a84 is .gnu.version_d in /lib/x86_64-linux-gnu/libdl.so.2
        0x00007f3e9f843a88 - 0x00007f3e9f843ae8 is .gnu.version_r in /lib/x86_64-linux-gnu/libdl.so.2
        0x00007f3e9f843ae8 - 0x00007f3e9f843c80 is .rela.dyn in /lib/x86_64-linux-gnu/libdl.so.2
        0x00007f3e9f843c80 - 0x00007f3e9f843db8 is .rela.plt in /lib/x86_64-linux-gnu/libdl.so.2
        0x00007f3e9f844000 - 0x00007f3e9f844017 is .init in /lib/x86_64-linux-gnu/libdl.so.2
        0x00007f3e9f844020 - 0x00007f3e9f844100 is .plt in /lib/x86_64-linux-gnu/libdl.so.2
        0x00007f3e9f844100 - 0x00007f3e9f844128 is .plt.got in /lib/x86_64-linux-gnu/libdl.so.2
        0x00007f3e9f844130 - 0x00007f3e9f844eb5 is .text in /lib/x86_64-linux-gnu/libdl.so.2
        0x00007f3e9f844eb8 - 0x00007f3e9f844ec1 is .fini in /lib/x86_64-linux-gnu/libdl.so.2
        0x00007f3e9f845000 - 0x00007f3e9f8450a3 is .rodata in /lib/x86_64-linux-gnu/libdl.so.2
--Type <RET> for more, q to quit, c to continue without paging--
        0x00007f3e9f8450a4 - 0x00007f3e9f845178 is .eh_frame_hdr in /lib/x86_64-linux-gnu/libdl.so.2
        0x00007f3e9f845178 - 0x00007f3e9f8454f0 is .eh_frame in /lib/x86_64-linux-gnu/libdl.so.2
        0x00007f3e9f8454f0 - 0x00007f3e9f8456a8 is .hash in /lib/x86_64-linux-gnu/libdl.so.2
        0x00007f3e9f846d70 - 0x00007f3e9f846d80 is .init_array in /lib/x86_64-linux-gnu/libdl.so.2
        0x00007f3e9f846d80 - 0x00007f3e9f846d90 is .fini_array in /lib/x86_64-linux-gnu/libdl.so.2
        0x00007f3e9f846d90 - 0x00007f3e9f846fa0 is .dynamic in /lib/x86_64-linux-gnu/libdl.so.2
        0x00007f3e9f846fa0 - 0x00007f3e9f847000 is .got in /lib/x86_64-linux-gnu/libdl.so.2
        0x00007f3e9f847000 - 0x00007f3e9f847080 is .got.plt in /lib/x86_64-linux-gnu/libdl.so.2
        0x00007f3e9f847080 - 0x00007f3e9f847088 is .data in /lib/x86_64-linux-gnu/libdl.so.2
        0x00007f3e9f8470a0 - 0x00007f3e9f847110 is .bss in /lib/x86_64-linux-gnu/libdl.so.2
        0x00007f3e9f6832e0 - 0x00007f3e9f683304 is .note.gnu.build-id in /lib/x86_64-linux-gnu/libc.so.6
        0x00007f3e9f683304 - 0x00007f3e9f683324 is .note.ABI-tag in /lib/x86_64-linux-gnu/libc.so.6
        0x00007f3e9f683328 - 0x00007f3e9f686fb0 is .gnu.hash in /lib/x86_64-linux-gnu/libc.so.6
        0x00007f3e9f686fb0 - 0x00007f3e9f694cd8 is .dynsym in /lib/x86_64-linux-gnu/libc.so.6
        0x00007f3e9f694cd8 - 0x00007f3e9f69ad16 is .dynstr in /lib/x86_64-linux-gnu/libc.so.6
        0x00007f3e9f69ad16 - 0x00007f3e9f69bf84 is .gnu.version in /lib/x86_64-linux-gnu/libc.so.6
        0x00007f3e9f69bf88 - 0x00007f3e9f69c3b0 is .gnu.version_d in /lib/x86_64-linux-gnu/libc.so.6
        0x00007f3e9f69c3b0 - 0x00007f3e9f69c3e0 is .gnu.version_r in /lib/x86_64-linux-gnu/libc.so.6
        0x00007f3e9f69c3e0 - 0x00007f3e9f6a3fd0 is .rela.dyn in /lib/x86_64-linux-gnu/libc.so.6
        0x00007f3e9f6a3fd0 - 0x00007f3e9f6a4438 is .rela.plt in /lib/x86_64-linux-gnu/libc.so.6
        0x00007f3e9f6a5000 - 0x00007f3e9f6a5300 is .plt in /lib/x86_64-linux-gnu/libc.so.6
        0x00007f3e9f6a5300 - 0x00007f3e9f6a5320 is .plt.got in /lib/x86_64-linux-gnu/libc.so.6
        0x00007f3e9f6a5320 - 0x00007f3e9f7eb14b is .text in /lib/x86_64-linux-gnu/libc.so.6
        0x00007f3e9f7eb150 - 0x00007f3e9f7ebf88 is __libc_freeres_fn in /lib/x86_64-linux-gnu/libc.so.6
        0x00007f3e9f7ec000 - 0x00007f3e9f80d3a8 is .rodata in /lib/x86_64-linux-gnu/libc.so.6
        0x00007f3e9f80d3b0 - 0x00007f3e9f80d3cc is .interp in /lib/x86_64-linux-gnu/libc.so.6
        0x00007f3e9f80d3cc - 0x00007f3e9f8135a8 is .eh_frame_hdr in /lib/x86_64-linux-gnu/libc.so.6
        0x00007f3e9f8135a8 - 0x00007f3e9f8340b8 is .eh_frame in /lib/x86_64-linux-gnu/libc.so.6
        0x00007f3e9f8340b8 - 0x00007f3e9f834588 is .gcc_except_table in /lib/x86_64-linux-gnu/libc.so.6
        0x00007f3e9f834588 - 0x00007f3e9f837a50 is .hash in /lib/x86_64-linux-gnu/libc.so.6
        0x00007f3e9f839648 - 0x00007f3e9f839658 is .tdata in /lib/x86_64-linux-gnu/libc.so.6
        0x00007f3e9f839658 - 0x00007f3e9f8396d8 is .tbss in /lib/x86_64-linux-gnu/libc.so.6
        0x00007f3e9f839658 - 0x00007f3e9f839668 is .init_array in /lib/x86_64-linux-gnu/libc.so.6
        0x00007f3e9f839668 - 0x00007f3e9f839758 is __libc_subfreeres in /lib/x86_64-linux-gnu/libc.so.6
        0x00007f3e9f839758 - 0x00007f3e9f839760 is __libc_atexit in /lib/x86_64-linux-gnu/libc.so.6
        0x00007f3e9f839760 - 0x00007f3e9f83a4c8 is __libc_IO_vtables in /lib/x86_64-linux-gnu/libc.so.6
        0x00007f3e9f83a4e0 - 0x00007f3e9f83cb80 is .data.rel.ro in /lib/x86_64-linux-gnu/libc.so.6
        0x00007f3e9f83cb80 - 0x00007f3e9f83cd60 is .dynamic in /lib/x86_64-linux-gnu/libc.so.6
        0x00007f3e9f83cd60 - 0x00007f3e9f83cff8 is .got in /lib/x86_64-linux-gnu/libc.so.6
        0x00007f3e9f83d000 - 0x00007f3e9f83d190 is .got.plt in /lib/x86_64-linux-gnu/libc.so.6
        0x00007f3e9f83d1a0 - 0x00007f3e9f83e860 is .data in /lib/x86_64-linux-gnu/libc.so.6
        0x00007f3e9f83e860 - 0x00007f3e9f842800 is .bss in /lib/x86_64-linux-gnu/libc.so.6
        0x00007f3e9f883238 - 0x00007f3e9f88325c is .note.gnu.build-id in /lib64/ld-linux-x86-64.so.2
        0x00007f3e9f883260 - 0x00007f3e9f883334 is .hash in /lib64/ld-linux-x86-64.so.2
        0x00007f3e9f883338 - 0x00007f3e9f883430 is .gnu.hash in /lib64/ld-linux-x86-64.so.2
--Type <RET> for more, q to quit, c to continue without paging--
        0x00007f3e9f883430 - 0x00007f3e9f883760 is .dynsym in /lib64/ld-linux-x86-64.so.2
        0x00007f3e9f883760 - 0x00007f3e9f883984 is .dynstr in /lib64/ld-linux-x86-64.so.2
        0x00007f3e9f883984 - 0x00007f3e9f8839c8 is .gnu.version in /lib64/ld-linux-x86-64.so.2
        0x00007f3e9f8839c8 - 0x00007f3e9f883a6c is .gnu.version_d in /lib64/ld-linux-x86-64.so.2
        0x00007f3e9f883a70 - 0x00007f3e9f883e60 is .rela.dyn in /lib64/ld-linux-x86-64.so.2
        0x00007f3e9f883e60 - 0x00007f3e9f883f08 is .rela.plt in /lib64/ld-linux-x86-64.so.2
        0x00007f3e9f884000 - 0x00007f3e9f884080 is .plt in /lib64/ld-linux-x86-64.so.2
        0x00007f3e9f884080 - 0x00007f3e9f884088 is .plt.got in /lib64/ld-linux-x86-64.so.2
        0x00007f3e9f884090 - 0x00007f3e9f8a1b50 is .text in /lib64/ld-linux-x86-64.so.2
```

```
0x00007f3e9f8a2000 - 0x00007f3e9f8a6620 is .rodata in /lib64/ld-linux-x86-64.so.2
0x00007f3e9f8a6620 - 0x00007f3e9f8a6cf4 is .eh_frame_hdr in /lib64/ld-linux-x86-64.so.2
0x00007f3e9f8a6cf8 - 0x00007f3e9f8a93dc is .eh_frame in /lib64/ld-linux-x86-64.so.2
0x00007f3e9f8aa640 - 0x00007f3e9f8aae74 is .data.rel.ro in /lib64/ld-linux-x86-64.so.2
0x00007f3e9f8aae78 - 0x00007f3e9f8aafe8 is .dynamic in /lib64/ld-linux-x86-64.so.2
0x00007f3e9f8aafe8 - 0x00007f3e9f8aaff8 is .got in /lib64/ld-linux-x86-64.so.2
0x00007f3e9f8ab000 - 0x00007f3e9f8ab050 is .got.plt in /lib64/ld-linux-x86-64.so.2
0x00007f3e9f8ab060 - 0x00007f3e9f8abff8 is .data in /lib64/ld-linux-x86-64.so.2
0x00007f3e9f8ac000 - 0x00007f3e9f8ac190 is .bss in /lib64/ld-linux-x86-64.so.2
0x00007f3e9f37b238 - 0x00007f3e9f37b25c is .note.gnu.build-id in /lib/x86_64-linux-gnu/libnss_files.so.2
0x00007f3e9f37b25c - 0x00007f3e9f37b27c is .note.ABI-tag in /lib/x86_64-linux-gnu/libnss_files.so.2
0x00007f3e9f37b280 - 0x00007f3e9f37b590 is .gnu.hash in /lib/x86_64-linux-gnu/libnss_files.so.2
0x00007f3e9f37b590 - 0x00007f3e9f37c0e8 is .dynsym in /lib/x86_64-linux-gnu/libnss_files.so.2
0x00007f3e9f37c0e8 - 0x00007f3e9f37ca5a is .dynstr in /lib/x86_64-linux-gnu/libnss_files.so.2
0x00007f3e9f37ca5a - 0x00007f3e9f37cb4c is .gnu.version in /lib/x86_64-linux-gnu/libnss_files.so.2
0x00007f3e9f37cb50 - 0x00007f3e9f37cb88 is .gnu.version_d in /lib/x86_64-linux-gnu/libnss_files.so.2
0x00007f3e9f37cb88 - 0x00007f3e9f37cbe8 is .gnu.version_r in /lib/x86_64-linux-gnu/libnss_files.so.2
0x00007f3e9f37cbe8 - 0x00007f3e9f37cd08 is .rela.dyn in /lib/x86_64-linux-gnu/libnss_files.so.2
0x00007f3e9f37cd08 - 0x00007f3e9f37d110 is .rela.plt in /lib/x86_64-linux-gnu/libnss_files.so.2
0x00007f3e9f37e000 - 0x00007f3e9f37e017 is .init in /lib/x86_64-linux-gnu/libnss_files.so.2
0x00007f3e9f37e020 - 0x00007f3e9f37e2e0 is .plt in /lib/x86_64-linux-gnu/libnss_files.so.2
0x00007f3e9f37e2e0 - 0x00007f3e9f37e2f8 is .plt.got in /lib/x86_64-linux-gnu/libnss_files.so.2
0x00007f3e9f37e300 - 0x00007f3e9f384578 is .text in /lib/x86_64-linux-gnu/libnss_files.so.2
0x00007f3e9f384578 - 0x00007f3e9f384581 is .fini in /lib/x86_64-linux-gnu/libnss_files.so.2
0x00007f3e9f385000 - 0x00007f3e9f385200 is .rodata in /lib/x86_64-linux-gnu/libnss_files.so.2
0x00007f3e9f385200 - 0x00007f3e9f38550c is .eh_frame_hdr in /lib/x86_64-linux-gnu/libnss_files.so.2
0x00007f3e9f385510 - 0x00007f3e9f386a90 is .eh_frame in /lib/x86_64-linux-gnu/libnss_files.so.2
0x00007f3e9f386a90 - 0x00007f3e9f386fe0 is .hash in /lib/x86_64-linux-gnu/libnss_files.so.2
0x00007f3e9f388d98 - 0x00007f3e9f388da0 is .init_array in /lib/x86_64-linux-gnu/libnss_files.so.2
0x00007f3e9f388da0 - 0x00007f3e9f388da8 is .fini_array in /lib/x86_64-linux-gnu/libnss_files.so.2
0x00007f3e9f388da8 - 0x00007f3e9f388fb8 is .dynamic in /lib/x86_64-linux-gnu/libnss_files.so.2
0x00007f3e9f388fb8 - 0x00007f3e9f389000 is .got in /lib/x86_64-linux-gnu/libnss_files.so.2
0x00007f3e9f389000 - 0x00007f3e9f389170 is .got.plt in /lib/x86_64-linux-gnu/libnss_files.so.2
0x00007f3e9f389170 - 0x00007f3e9f389178 is .data in /lib/x86_64-linux-gnu/libnss_files.so.2
0x00007f3e9f389180 - 0x00007f3e9f38f738 is .bss in /lib/x86_64-linux-gnu/libnss_files.so.2
0x00007ffcbd7ea120 - 0x00007ffcbd7ea164 is .hash in system-supplied DSO at 0x7ffcbd7ea000
--Type <RET> for more, q to quit, c to continue without paging--
0x00007ffcbd7ea168 - 0x00007ffcbd7ea1b8 is .gnu.hash in system-supplied DSO at 0x7ffcbd7ea000
0x00007ffcbd7ea1b8 - 0x00007ffcbd7ea2d8 is .dynsym in system-supplied DSO at 0x7ffcbd7ea000
0x00007ffcbd7ea2d8 - 0x00007ffcbd7ea34a is .dynstr in system-supplied DSO at 0x7ffcbd7ea000
0x00007ffcbd7ea34a - 0x00007ffcbd7ea362 is .gnu.version in system-supplied DSO at 0x7ffcbd7ea000
0x00007ffcbd7ea368 - 0x00007ffcbd7ea3a0 is .gnu.version_d in system-supplied DSO at 0x7ffcbd7ea000
0x00007ffcbd7ea3a0 - 0x00007ffcbd7ea4b0 is .dynamic in system-supplied DSO at 0x7ffcbd7ea000
0x00007ffcbd7ea4b0 - 0x00007ffcbd7ea504 is .note in system-supplied DSO at 0x7ffcbd7ea000
0x00007ffcbd7ea504 - 0x00007ffcbd7ea538 is .eh_frame_hdr in system-supplied DSO at 0x7ffcbd7ea000
0x00007ffcbd7ea538 - 0x00007ffcbd7ea5fc is .eh_frame in system-supplied DSO at 0x7ffcbd7ea000
0x00007ffcbd7ea600 - 0x00007ffcbd7ea915 is .text in system-supplied DSO at 0x7ffcbd7ea000
0x00007ffcbd7ea915 - 0x00007ffcbd7ea98a is .altinstructions in system-supplied DSO at 0x7ffcbd7ea000
0x00007ffcbd7ea98a - 0x00007ffcbd7ea9ae is .altinstr_replacement in system-supplied DSO at 0x7ffcbd7ea000
```

7.   Find shared libraries that are dynamically linked with *free* functions and verify that they all use the same *free* function:

```
(gdb) info functions free@plt
All functions matching regular expression "free@plt":

Non-debugging symbols:
0x000055e165078080  free@plt
0x000055e165078370  regfree@plt
0x00007f3e9f856070  free@plt
0x00007f3e9f844040  free@plt
0x00007f3e9f6a5318  free@plt
0x00007f3e9f884080  free@plt
0x00007f3e9f37e050  free@plt
```

Note: We check 5 entries except the one that we've already done.

```
(gdb) disassemble 0x000055e165078080
Dump of assembler code for function free@plt:
    0x000055e165078080 <+0>:     jmpq    *0xe7fba(%rip)        # 0x55e165160040 <free@got.plt>
    0x000055e165078086 <+6>:     pushq   $0x5
    0x000055e16507808b <+11>:    jmpq    0x55e165078020
End of assembler dump.

(gdb) x/a 0x55e165160040
0x55e165160040 <free@got.plt>:  0x7f3e9f707b60 <__GI___libc_free>

(gdb) disassemble 0x00007f3e9f856070
Dump of assembler code for function free@plt:
    0x00007f3e9f856070 <+0>:     jmpq    *0x1ea1a(%rip)        # 0x7f3e9f874a90 <free@got.plt>
    0x00007f3e9f856076 <+6>:     pushq   $0x4
    0x00007f3e9f85607b <+11>:    jmpq    0x7f3e9f856020
End of assembler dump.

(gdb) x/a 0x7f3e9f874a90
0x7f3e9f874a90 <free@got.plt>:  0x7f3e9f707b60 <__GI___libc_free>

(gdb) disassemble 0x00007f3e9f844040
Dump of assembler code for function free@plt:
    0x00007f3e9f844040 <+0>:     jmpq    *0x2fda(%rip)        # 0x7f3e9f847020 <free@got.plt>
    0x00007f3e9f844046 <+6>:     pushq   $0x1
    0x00007f3e9f84404b <+11>:    jmpq    0x7f3e9f844020
End of assembler dump.

(gdb) x/a 0x7f3e9f847020
0x7f3e9f847020 <free@got.plt>:  0x7f3e9f844046 <free@plt+6>

(gdb) disassemble 0x00007f3e9f6a5318
Dump of assembler code for function free@plt:
    0x00007f3e9f6a5318 <+0>:     jmpq    *0x197c8a(%rip)        # 0x7f3e9f83cfa8
    0x00007f3e9f6a531e <+6>:     xchg    %ax,%ax
End of assembler dump.

(gdb) x/a 0x7f3e9f83cfa8
0x7f3e9f83cfa8: 0x7f3e9f707b60 <__GI___libc_free>

(gdb) disassemble 0x00007f3e9f37e050
Dump of assembler code for function free@plt:
    0x00007f3e9f37e050 <+0>:     jmpq    *0xafd2(%rip)        # 0x7f3e9f389028 <free@got.plt>
    0x00007f3e9f37e056 <+6>:     pushq   $0x2
    0x00007f3e9f37e05b <+11>:    jmpq    0x7f3e9f37e020
End of assembler dump.

(gdb) x/a 0x7f3e9f389028
0x7f3e9f389028 <free@got.plt>:  0x7f3e9f707b60 <__GI___libc_free>
```

Note: Checking these entries with the **info files** command output above, we see the following shared libraries in addition to *bash*:

```
0x00007f3e9f6a5000 - 0x00007f3e9f6a5300 is .plt in /lib/x86_64-linux-gnu/libc.so.6

0x00007f3e9f856020 - 0x00007f3e9f856940 is .plt in /lib/x86_64-linux-gnu/libtinfo.so.6

0x00007f3e9f37e020 - 0x00007f3e9f37e2e0 is .plt in /lib/x86_64-linux-gnu/libnss_files.so.2

0x00007f3e9f844020 - 0x00007f3e9f844100 is .plt in /lib/x86_64-linux-gnu/libdl.so.2
```

8.   We now check what functions the *libtinfo.so.6* share library imports:

```
0x00007f3e9f856020 - 0x00007f3e9f856940 is .plt in /lib/x86_64-linux-gnu/libtinfo.so.6
```

```
(gdb) p (0x00007f3e9f856940-0x00007f3e9f856020)/8
$1 = 292
```

```
(gdb) x/292a 0x00007f3e9f856020
0x7f3e9f856020: 0x25ff0001ea3a35ff          0x401f0f0001ea3c
0x7f3e9f856030 <_nc_export_termtype2@plt>:        0x680001ea3a25ff      0xfffffffe0e9000000
0x7f3e9f856040 <getenv@plt>:        0x1680001ea3225ff        0xfffffffd0e9000000
0x7f3e9f856050 <noraw_sp@plt>:  0x2680001ea2a25ff        0xfffffffc0e9000000
0x7f3e9f856060 <_nc_get_tty_mode_sp@plt>:        0x3680001ea2225ff        0xfffffffb0e9000000
0x7f3e9f856070 <free@plt>:        0x4680001ea1a25ff        0xfffffffa0e9000000
0x7f3e9f856080 <_nc_setupterm@plt>:        0x5680001ea1225ff        0xfffffff90e9000000
0x7f3e9f856090 <_nc_pathlast@plt>:        0x6680001ea0a25ff        0xfffffff80e9000000
0x7f3e9f8560a0 <killchar_sp@plt>:        0x7680001ea0225ff        0xfffffff70e9000000
0x7f3e9f8560b0 <__vfprintf_chk@plt>:        0x8680001e9fa25ff        0xfffffff60e9000000
0x7f3e9f8560c0 <__errno_location@plt>:  0x9680001e9f225ff        0xfffffff50e9000000
0x7f3e9f8560d0 <strncpy@plt>:        0xa680001e9ea25ff        0xfffffff40e9000000
0x7f3e9f8560e0 <strncmp@plt>:        0xb680001e9e225ff        0xfffffff30e9000000
0x7f3e9f8560f0 <keybound_sp@plt>:        0xc680001e9da25ff        0xfffffff20e9000000
0x7f3e9f856100 <strcpy@plt>:        0xd680001e9d225ff        0xfffffff10e9000000
0x7f3e9f856110 <_nc_set_no_padding@plt>:        0xe680001e9ca25ff        0xfffffff00e9000000
0x7f3e9f856120 <_nc_getenv_num@plt>:        0xf680001e9c225ff        0xffffffef0e9000000
0x7f3e9f856130 <del_curterm_sp@plt>:        0x10680001e9ba25ff        0xffffffee0e9000000
0x7f3e9f856140 <tgetstr_sp@plt>:        0x11680001e9b225ff        0xffffffed0e9000000
0x7f3e9f856150 <isatty@plt>:        0x12680001e9aa25ff        0xffffffec0e9000000
0x7f3e9f856160 <fread@plt>:        0x13680001e9a225ff        0xffffffeb0e9000000
0x7f3e9f856170 <baudrate@plt>:  0x14680001e99a25ff        0xffffffea0e9000000
0x7f3e9f856180 <reset_shell_mode_sp@plt>:        0x15680001e99225ff        0xffffffe90e9000000
0x7f3e9f856190 <has_ic_sp@plt>: 0x16680001e98a25ff        0xffffffe80e9000000
0x7f3e9f8561a0 <setenv@plt>:        0x17680001e98225ff        0xffffffe70e9000000
0x7f3e9f8561b0 <write@plt>:        0x18680001e97a25ff        0xffffffe60e9000000
0x7f3e9f8561c0 <set_tabsize_sp@plt>:        0x19680001e97225ff        0xffffffe50e9000000
0x7f3e9f8561d0 <tgetent_sp@plt>:        0x1a680001e96a25ff        0xffffffe40e9000000
0x7f3e9f8561e0 <resetty_sp@plt>:        0x1b680001e96225ff        0xffffffe30e9000000
0x7f3e9f8561f0 <has_il_sp@plt>: 0x1c680001e95a25ff        0xffffffe20e9000000
0x7f3e9f856200 <fclose@plt>:        0x1d680001e95225ff        0xffffffe10e9000000
0x7f3e9f856210 <tigetnum_sp@plt>:        0x1e680001e94a25ff        0xffffffe00e9000000
0x7f3e9f856220 <_nc_doalloc@plt>:        0x1f680001e94225ff        0xffffffdf0e9000000
0x7f3e9f856230 <flushinp_sp@plt>:        0x20680001e93a25ff        0xffffffde0e9000000
0x7f3e9f856240 <strlen@plt>:        0x21680001e93225ff        0xffffffdd0e9000000
0x7f3e9f856250 <erasechar_sp@plt>:        0x22680001e92a25ff        0xffffffdc0e9000000
0x7f3e9f856260 <tputs_sp@plt>:  0x23680001e92225ff        0xffffffdb0e9000000
0x7f3e9f856270 <def_prog_mode_sp@plt>:  0x24680001e91a25ff        0xffffffda0e9000000
0x7f3e9f856280 <tigetstr_sp@plt>:        0x25680001e91225ff        0xffffffd90e9000000
0x7f3e9f856290 <_nc_access@plt>:        0x26680001e90a25ff        0xffffffd80e9000000
0x7f3e9f8562a0 <__stack_chk_fail@plt>:        0x27680001e90225ff        0xffffffd70e9000000
0x7f3e9f8562b0 <has_key_sp@plt>:        0x28680001e8fa25ff        0xffffffd60e9000000
0x7f3e9f8562c0 <tcflush@plt>:        0x29680001e8f225ff        0xffffffd50e9000000
0x7f3e9f8562d0 <strchr@plt>:        0x2a680001e8ea25ff        0xffffffd40e9000000
0x7f3e9f8562e0 <keyname_sp@plt>:        0x2b680001e8e225ff        0xffffffd30e9000000
--Type <RET> for more, q to quit, c to continue without paging--
0x7f3e9f8562f0 <key_defined_sp@plt>:        0x2c680001e8da25ff        0xffffffd20e9000000
0x7f3e9f856300 <_nc_keypad@plt>:        0x2d680001e8d225ff        0xffffffd10e9000000
0x7f3e9f856310 <reset_prog_mode_sp@plt>:        0x2e680001e8ca25ff        0xffffffd00e9000000
0x7f3e9f856320 <nanosleep@plt>: 0x2f680001e8c225ff        0xffffffcf0e9000000
0x7f3e9f856330 <strrchr@plt>:        0x30680001e8ba25ff        0xffffffce0e9000000
0x7f3e9f856340 <_nc_read_file_entry@plt>:        0x31680001e8b225ff        0xffffffcd0e9000000
```

117

```
0x7f3e9f856350 <_nc_get_screensize@plt>:        0x32680001e8aa25ff      0xffffffcc0e9000000
0x7f3e9f856360 <tparm@plt>:        0x33680001e8a225ff      0xffffffcb0e9000000
0x7f3e9f856370 <gettimeofday@plt>:        0x34680001e89a25ff      0xffffffca0e9000000
0x7f3e9f856380 <def_shell_mode_sp@plt>: 0x35680001e89225ff      0xffffffc90e9000000
0x7f3e9f856390 <_nc_get_table@plt>:        0x36680001e88a25ff      0xffffffc80e9000000
0x7f3e9f8563a0 <memset@plt>:        0x37680001e88225ff      0xffffffc70e9000000
0x7f3e9f8563b0 <termname_sp@plt>:        0x38680001e87a25ff      0xffffffc60e9000000
0x7f3e9f8563c0 <_nc_str_init@plt>:        0x39680001e87225ff      0xffffffc50e9000000
0x7f3e9f8563d0 <__poll_chk@plt>:        0x3a680001e86a25ff      0xffffffc40e9000000
0x7f3e9f8563e0 <ioctl@plt>:        0x3b680001e86225ff      0xffffffc30e9000000
0x7f3e9f8563f0 <_nc_tparm_analyze@plt>: 0x3c680001e85a25ff      0xffffffc20e9000000
0x7f3e9f856400 <_nc_free_termtype@plt>: 0x3d680001e85225ff      0xffffffc10e9000000
0x7f3e9f856410 <strncat@plt>:        0x3e680001e84a25ff      0xffffffc00e9000000
0x7f3e9f856420 <putp_sp@plt>:        0x3f680001e84225ff      0xffffffbf0e9000000
0x7f3e9f856430 <baudrate_sp@plt>:        0x40680001e83a25ff      0xffffffbe0e9000000
0x7f3e9f856440 <fputc@plt>:        0x41680001e83225ff      0xffffffbd0e9000000
0x7f3e9f856450 <_nc_find_type_entry@plt>:        0x42680001e82a25ff      0xffffffbc0e9000000
0x7f3e9f856460 <calloc@plt>:        0x43680001e82225ff      0xffffffbb0e9000000
0x7f3e9f856470 <strcmp@plt>:        0x44680001e81a25ff      0xffffffba0e9000000
0x7f3e9f856480 <putc@plt>:        0x45680001e81225ff      0xffffffb90e9000000
0x7f3e9f856490 <set_curterm@plt>:        0x46680001e80a25ff      0xffffffb80e9000000
0x7f3e9f8564a0 <_nc_next_db@plt>:        0x47680001e80225ff      0xffffffb70e9000000
0x7f3e9f8564b0 <__memcpy_chk@plt>:        0x48680001e7fa25ff      0xffffffb60e9000000
0x7f3e9f8564c0 <tgetflag_sp@plt>:        0x49680001e7f225ff      0xffffffb50e9000000
0x7f3e9f8564d0 <strtol@plt>:        0x4a680001e7ea25ff      0xffffffb40e9000000
0x7f3e9f8564e0 <_nc_set_tty_mode@plt>: 0x4b680001e7e225ff      0xffffffb30e9000000
0x7f3e9f8564f0 <memcpy@plt>:        0x4c680001e7da25ff      0xffffffb20e9000000
0x7f3e9f856500 <_nc_putp_sp@plt>:        0x4d680001e7d225ff      0xffffffb10e9000000
0x7f3e9f856510 <time@plt>:        0x4e680001e7ca25ff      0xffffffb00e9000000
0x7f3e9f856520 <fileno@plt>:        0x4f680001e7c225ff      0xffffffaf0e9000000
0x7f3e9f856530 <intrflush_sp@plt>:        0x50680001e7ba25ff      0xffffffae0e9000000
0x7f3e9f856540 <_nc_init_acs_sp@plt>:        0x51680001e7b225ff      0xffffffad0e9000000
0x7f3e9f856550 <__xstat@plt>:        0x52680001e7aa25ff      0xffffffac0e9000000
0x7f3e9f856560 <cbreak_sp@plt>: 0x53680001e7a225ff      0xffffffab0e9000000
0x7f3e9f856570 <_nc_tic_dir@plt>:        0x54680001e79a25ff      0xffffffaa0e9000000
0x7f3e9f856580 <malloc@plt>:        0x55680001e79225ff      0xffffffa90e9000000
0x7f3e9f856590 <unctrl_sp@plt>: 0x56680001e78a25ff      0xffffffa80e9000000
0x7f3e9f8565a0 <fflush@plt>:        0x57680001e78225ff      0xffffffa70e9000000
0x7f3e9f8565b0 <nl_langinfo@plt>:        0x58680001e77a25ff      0xffffffa60e9000000
--Type <RET> for more, q to quit, c to continue without paging--
0x7f3e9f8565c0 <savetty_sp@plt>:        0x59680001e77225ff      0xffffffa50e9000000
0x7f3e9f8565d0 <longname@plt>: 0x5a680001e76a25ff      0xffffffa40e9000000
0x7f3e9f8565e0 <nocbreak_sp@plt>:        0x5b680001e76225ff      0xffffffa30e9000000
0x7f3e9f8565f0 <_nc_fallback2@plt>:        0x5c680001e75a25ff      0xffffffa20e9000000
0x7f3e9f856600 <noqiflush_sp@plt>:        0x5d680001e75225ff      0xffffffa10e9000000
0x7f3e9f856610 <qiflush_sp@plt>:        0x5e680001e74a25ff      0xffffffa00e9000000
0x7f3e9f856620 <del_curterm@plt>:        0x5f680001e74225ff      0xffffff9f0e9000000
0x7f3e9f856630 <_nc_warning@plt>:        0x60680001e73a25ff      0xffffff9e0e9000000
0x7f3e9f856640 <realloc@plt>:        0x61680001e73225ff      0xffffff9d0e9000000
0x7f3e9f856650 <_nc_flush@plt>: 0x62680001e72a25ff      0xffffff9c0e9000000
0x7f3e9f856660 <setlocale@plt>: 0x63680001e72225ff      0xffffff9b0e9000000
0x7f3e9f856670 <delay_output_sp@plt>:        0x64680001e71a25ff      0xffffff9a0e9000000
0x7f3e9f856680 <cfgetospeed@plt>:        0x65680001e71225ff      0xffffff990e9000000
0x7f3e9f856690 <_nc_err_abort@plt>:        0x66680001e70a25ff      0xffffff980e9000000
0x7f3e9f8566a0 <_nc_set_buffer_sp@plt>: 0x67680001e70225ff      0xffffff970e9000000
0x7f3e9f8566b0 <_nc_putp_flush_sp@plt>: 0x68680001e6fa25ff      0xffffff960e9000000
0x7f3e9f8566c0 <_nc_add_to_try@plt>:        0x69680001e6f225ff      0xffffff950e9000000
0x7f3e9f8566d0 <_nc_basename@plt>:        0x6a680001e6ea25ff      0xffffff940e9000000
0x7f3e9f8566e0 <keyok_sp@plt>: 0x6b680001e6e225ff      0xffffff930e9000000
0x7f3e9f8566f0 <tcgetattr@plt>: 0x6c680001e6da25ff      0xffffff920e9000000
```

```
0x7f3e9f856700 <tcsetattr@plt>:      0x6d680001e6d225ff      0xfffff910e9000000
0x7f3e9f856710 <curs_set_sp@plt>:            0x6e680001e6ca25ff      0xfffff900e9000000
0x7f3e9f856720 <access@plt>:      0x6f680001e6c225ff      0xfffff8f0e9000000
0x7f3e9f856730 <fopen@plt>:          0x70680001e6ba25ff      0xfffff8e0e9000000
0x7f3e9f856740 <sysconf@plt>:        0x71680001e6b225ff      0xfffff8d0e9000000
0x7f3e9f856750 <_nc_name_match@plt>:         0x72680001e6aa25ff      0xfffff8c0e9000000
0x7f3e9f856760 <raw_sp@plt>:       0x73680001e6a225ff      0xfffff8b0e9000000
0x7f3e9f856770 <halfdelay_sp@plt>:           0x74680001e69a25ff      0xfffff8a0e9000000
0x7f3e9f856780 <_nc_set_tty_mode_sp@plt>:          0x75680001e69225ff      0xfffff890e9000000
0x7f3e9f856790 <_nc_get_hash_table@plt>:           0x76680001e68a25ff      0xfffff880e9000000
0x7f3e9f8567a0 <_nc_home_terminfo@plt>: 0x77680001e68225ff      0xfffff870e9000000
0x7f3e9f8567b0 <_nc_first_db@plt>:           0x78680001e67a25ff      0xfffff860e9000000
0x7f3e9f8567c0 <tigetflag_sp@plt>:           0x79680001e67225ff      0xfffff850e9000000
0x7f3e9f8567d0 <_nc_trim_sgr0@plt>:          0x7a680001e66a25ff      0xfffff840e9000000
0x7f3e9f8567e0 <exit@plt>:       0x7b680001e66225ff      0xfffff830e9000000
0x7f3e9f8567f0 <fwrite@plt>:         0x7c680001e65a25ff      0xfffff820e9000000
0x7f3e9f856800 <typeahead_sp@plt>:           0x7d680001e65225ff      0xfffff810e9000000
0x7f3e9f856810 <__fprintf_chk@plt>:          0x7e680001e64a25ff      0xfffff800e9000000
0x7f3e9f856820 <napms_sp@plt>:  0x7f680001e64225ff      0xfffff7f0e9000000
0x7f3e9f856830 <_nc_read_termtype@plt>: 0x80680001e63a25ff      0xfffff7e0e9000000
0x7f3e9f856840 <_nc_read_entry2@plt>:        0x81680001e63225ff      0xfffff7d0e9000000
0x7f3e9f856850 <napms@plt>:      0x82680001e62a25ff      0xfffff7c0e9000000
0x7f3e9f856860 <strdup@plt>:     0x83680001e62225ff      0xfffff7b0e9000000
0x7f3e9f856870 <set_curterm_sp@plt>:         0x84680001e61a25ff      0xfffff7a0e9000000
0x7f3e9f856880 <_nc_flush_sp@plt>:           0x85680001e61225ff      0xfffff790e9000000
--Type <RET> for more, q to quit, c to continue without paging--
0x7f3e9f856890 <define_key_sp@plt>:          0x86680001e60a25ff      0xfffff780e9000000
0x7f3e9f8568a0 <_nc_last_db@plt>:            0x87680001e60225ff      0xfffff770e9000000
0x7f3e9f8568b0 <tigetnum@plt>:  0x88680001e5fa25ff      0xfffff760e9000000
0x7f3e9f8568c0 <_nc_get_tty_mode@plt>: 0x89680001e5f225ff      0xfffff750e9000000
0x7f3e9f8568d0 <strstr@plt>:     0x8a680001e5ea25ff      0xfffff740e9000000
0x7f3e9f8568e0 <_nc_visbuf2@plt>:            0x8b680001e5e225ff      0xfffff730e9000000
0x7f3e9f8568f0 <tgetnum_sp@plt>:             0x8c680001e5da25ff      0xfffff720e9000000
0x7f3e9f856900 <__ctype_b_loc@plt>:          0x8d680001e5d225ff      0xfffff710e9000000
0x7f3e9f856910 <_nc_free_termtype2@plt>:          0x8e680001e5ca25ff          0xfffff700e9000000
0x7f3e9f856920 <_nc_screen_of@plt>:          0x8f680001e5c225ff      0xfffff6f0e9000000
0x7f3e9f856930 <__sprintf_chk@plt>:          0x90680001e5ba25ff      0xfffff6e0e9000000
```

Note: Similar information is also available from the .GOT section.

```
0x00007f3e9f874a58 - 0x00007f3e9f874ff8 is .got in /lib/x86_64-linux-gnu/libtinfo.so.6
```

```
(gdb) p (0x00007f3e9f874ff8-0x00007f3e9f874a58)/8
$2 = 180
```

```
(gdb) x/180a 0x00007f3e9f874a58
0x7f3e9f874a58: 0x2c848 0x0
0x7f3e9f874a68: 0x0      0x7f3e9f857f60 <_nc_copy_termtype2>
0x7f3e9f874a78 <getenv@got.plt>:             0x55e1650f87b0 <getenv> 0x7f3e9f85ae50 <qiflush>
0x7f3e9f874a88 <_nc_get_tty_mode_sp@got.plt>:   0x7f3e9f860da0 0x7f3e9f707b60 <__GI___libc_free>
0x7f3e9f874a98 <_nc_setupterm@got.plt>: 0x7f3e9f85ba70 <_nc_locale_breaks_acs+352>      0x7f3e9f856a10 <_nc_is_abs_path>
0x7f3e9f874aa8 <killchar_sp@got.plt>:   0x7f3e9f859f20 <erasechar>      0x7f3e9f85b250 <___vfprintf_chk>
0x7f3e9f874ab8 <__errno_location@got.plt>:      0x7f3e9f6a7330 <__GI___errno_location>  0x7f3e9f7205d0 <__strncpy_sse2_unaligned>
0x7f3e9f874ac8 <strncmp@got.plt>:       0x7f3e9f7da8a0 <__strncmp_avx2> 0x7f3e9f863b50
0x7f3e9f874ad8 <strcpy@got.plt>:        0x7f3e9f71ffa0 <__strcpy_sse2_unaligned>        0x7f3e9f860510 <_nc_outc_wrapper+32>
0x7f3e9f874ae8 <_nc_getenv_num@got.plt>:        0x7f3e9f859430 <use_extended_names>     0x7f3e9f859cc0 <set_curterm+16>
0x7f3e9f874af8 <tgetstr_sp@got.plt>:    0x7f3e9f85c980 <tgetnum+16>     0x7f3e9f76ed40 <__isatty>
0x7f3e9f874b08 <fread@got.plt>: 0x7f3e9f6f3720 <__GI__IO_fread> 0x7f3e9f859c00 <baudrate_sp+176>
0x7f3e9f874b18 <reset_shell_mode_sp@got.plt>:   0x7f3e9f861060 <reset_prog_mode>        0x7f3e9f859da0 <_nc_screen_of+16>
0x7f3e9f874b28 <setenv@got.plt>:        0x55e1650f8980 <setenv> 0x7f3e9f76d3a0 <__GI___libc_write>
0x7f3e9f874b38 <set_tabsize_sp@got.plt>:        0x7f3e9f85b3a0 <intrflush+16>   0x7f3e9f85c010 <setupterm>
0x7f3e9f874b48 <resetty_sp@got.plt>:    0x7f3e9f861110 <savetty>        0x7f3e9f859e40 <has_ic>
0x7f3e9f874b58 <fclose@got.plt>:        0x7f3e9f6f2910 <_IO_new_fclose> 0x7f3e9f85ccf0 <tigetflag+16>
0x7f3e9f874b68 <_nc_doalloc@got.plt>:   0x7f3e9f859200 <_nc_first_db+1520>      0x7f3e9f859f70 <killchar>
0x7f3e9f874b78 <strlen@got.plt>:        0x7f3e9f7def20 <__strlen_avx2>  0x7f3e9f859ed0 <has_il>
0x7f3e9f874b88 <tputs_sp@got.plt>:      0x7f3e9f860890 <_nc_putchar>    0x7f3e9f860f70 <def_shell_mode>
```

119

```
0x7f3e9f874b98 <tigetstr_sp@got.plt>:    0x7f3e9f85ce20 <tigetnum+16>    0x7f3e9f856a60 <_nc_rootname>
0x7f3e9f874ba8 <__stack_chk_fail@got.plt>:     0x7f3e9f78d350 <__stack_chk_fail>    0x7f3e9f85a780 <typeahead+16>
0x7f3e9f874bb8 <tcflush@got.plt>:    0x7f3e9f7728c0 <tcflush>    0x7f3e9f7de930 <__strchr_avx2>
0x7f3e9f874bc8 <keyname_sp@got.plt>:    0x7f3e9f859ff0 <flushinp>    0x7f3e9f863b00
0x7f3e9f874bd8 <_nc_keypad@got.plt>:    0x7f3e9f85a930 <curs_set+16>    0x7f3e9f860fd0 <def_prog_mode>
0x7f3e9f874be8 <nanosleep@got.plt>:    0x7f3e9f749580 <__GI___nanosleep>    0x7f3e9f7ded50 <__strrchr_avx2>
0x7f3e9f874bf8 <_nc_read_file_entry@got.plt>:    0x7f3e9f8625b0    0x7f3e9f85b470 <use_tioctl>
0x7f3e9f874c08 <tparm@got.plt>: 0x7f3e9f85d420 <_nc_tparm_analyze+1232>  0x7ffcbd7ea600 <gettimeofday>
0x7f3e9f874c18 <def_shell_mode_sp@got.plt>:    0x7f3e9f860ef0 <_nc_set_tty_mode+16>    0x7f3e9f858250
0x7f3e9f874c28 <memset@got.plt>:    0x7f3e9f7df8e0 <__memset_avx2_unaligned_erms>    0x7f3e9f85cb60 <tgetstr+16>
0x7f3e9f874c38 <_nc_str_init@got.plt>:    0x7f3e9f862b40 <_nc_read_entry+16>    0x7f3e9f78d2e0 <__poll_chk>
0x7f3e9f874c48 <ioctl@got.plt>: 0x7f3e9f772fc0 <ioctl>  0x7f3e9f85cf40 <tigetstr+16>
0x7f3e9f874c58 <_nc_free_termtype@got.plt>:    0x7f3e9f859400    0x7f3e9f723880 <__strncat_sse2_unaligned>
0x7f3e9f874c68 <putp_sp@got.plt>:    0x7f3e9f860c40 <tputs_sp+928>    0x7f3e9f859b40
0x7f3e9f874c78 <fputc@got.plt>: 0x7f3e9f6fa9f0 <fputc>  0x7f3e9f8588e0 <_nc_find_entry+160>
0x7f3e9f874c88 <calloc@got.plt>:    0x7f3e9f7082e0 <__libc_calloc>    0x7f3e9f7da460 <__strcmp_avx2>
0x7f3e9f874c98 <putc@got.plt>:  0x7f3e9f6fb2a0 <__GI__IO_putc>    0x7f3e9f859ca0 <set_curterm_sp+128>
0x7f3e9f874ca8 <_nc_next_db@got.plt>:    0x7f3e9f858bd0 <_nc_last_db+48>    0x7f3e9f7df450 <__memmove_chk_avx_unaligned_erms>
0x7f3e9f874cb8 <tgetflag_sp@got.plt>:    0x7f3e9f85c700 <tgetent+16>    0x7f3e9f6bed50 <__strtol>
0x7f3e9f874cc8 <_nc_set_tty_mode@got.plt>:    0x7f3e9f860ed0 <_nc_set_tty_mode_sp+112>    0x7f3e9f7df460
<__memmove_avx_unaligned_erms>
0x7f3e9f874cd8 <_nc_putp@got.plt>:    0x7f3e9f860c80 <putp+16>    0x7ffcbd7ea730 <time>
0x7f3e9f874ce8 <fileno@got.plt>:    0x7f3e9f6fa9c0 <__GI___fileno>    0x7f3e9f85b280 <noqiflush>
0x7f3e9f874cf8 <_nc_init_acs_sp@got.plt>:    0x7f3e9f859740    0x7f3e9f76c940 <__GI___xstat>
0x7f3e9f874d08 <cbreak_sp@got.plt>:    0x7f3e9f85aba0 <raw>    0x7f3e9f858b00
0x7f3e9f874d18 <malloc@got.plt>:    0x7f3e9f707510 <__GI___libc_malloc>    0x7f3e9f863580 <_nc_trim_sgr0+1120>
--Type <RET> for more, q to quit, c to continue without paging--
0x7f3e9f874d28 <fflush@got.plt>:    0x7f3e9f6f2dc0 <__GI__IO_fflush>    0x7f3e9f6b2140 <__GI_nl_langinfo>
0x7f3e9f874d38 <savetty_sp@got.plt>:    0x7f3e9f8610d0 <reset_shell_mode>    0x7f3e9f85a370 <keyname+16>
0x7f3e9f874d48 <nocbreak_sp@got.plt>:    0x7f3e9f85aff0 <noraw>    0x7f3e9f859320 <_nc_leaks_tinfo>
0x7f3e9f874d58 <noqiflush_sp@got.plt>:    0x7f3e9f85b180 <nocbreak>    0x7f3e9f85ad50 <cbreak>
0x7f3e9f874d68 <del_curterm@got.plt>:    0x7f3e9f859d60 <del_curterm_sp+144>    0x7f3e9f707db0 <__GI___libc_realloc>
0x7f3e9f874d78 <realloc@got.plt>:    0x7f3e9f860850 <_nc_outch_sp+240>    0x7f3e9f6b0520 <__GI_setlocale>
0x7f3e9f874d88 <setlocale@got.plt>:    0x7f3e9f860600 <_nc_flush_sp+160>    0x7f3e9f7723a0 <cfgetospeed>
0x7f3e9f874d98 <cfgetospeed@got.plt>:    0x7f3e9f858690 <_nc_warning+432>    0x7f3e9f8614a0 <_nc_name_match+192>
0x7f3e9f874da8 <_nc_set_buffer_sp@got.plt>:    0x7f3e9f85a7d0 <has_key+16>    0x7f3e9f856c20 <_nc_is_file_path+96>
0x7f3e9f874db8 <_nc_add_to_try@got.plt>:    0x7f3e9f856a40 <_nc_pathlast+32>    0x7f3e9f863bb0 <key_defined+16>
0x7f3e9f874dc8 <keyok_sp@got.plt>:    0x7f3e9f7726d0 <__GI___tcgetattr>    0x7f3e9f7724f0 <__tcsetattr>
0x7f3e9f874dd8 <tcsetattr@got.plt>:    0x7f3e9f85a7f0 <_nc_putp_flush_sp+16>    0x7f3e9f76d470 <__access>
0x7f3e9f874de8 <access@got.plt>:    0x7f3e9f6f3310 <_IO_new_fopen>    0x7f3e9f74b230 <__GI___sysconf>
0x7f3e9f874df8 <sysconf@got.plt>:    0x7f3e9f8613d0 <_nc_first_name+128>    0x7f3e9f85a9f0 <keypad+32>
0x7f3e9f874e08 <raw_sp@got.plt>:    0x7f3e9f85a5c0 <idcok+80>    0x7f3e9f860e50 <_nc_get_tty_mode+16>
0x7f3e9f874e18 <_nc_set_tty_mode_sp@got.plt>:    0x7f3e9f858290 <_nc_get_table+48>    0x7f3e9f859550
0x7f3e9f874e28 <_nc_home_terminfo@got.plt>:    0x7f3e9f858c00 <_nc_next_db+32>    0x7f3e9f85cbd0 <tgoto>
0x7f3e9f874e38 <tigetflag@got.plt>:    0x7f3e9f8630b0    0x7f3e9f6bcfd0 <__GI_exit>
0x7f3e9f874e48 <exit@got.plt>:  0x7f3e9f6f3b40 <__GI__IO_fwrite>    0x7f3e9f85a730 <meta+128>
0x7f3e9f874e58 <typeahead_sp@got.plt>:    0x7f3e9f78afd0 <___fprintf_chk>    0x7f3e9f85a3c0 <longname_sp>
0x7f3e9f874e68 <napms_sp@got.plt>:    0x7f3e9f861770 <_nc_init_termtype+176>    0x7f3e9f862970
0x7f3e9f874e78 <_nc_read_entry2@got.plt>:    0x7f3e9f85a460 <napms_sp+144>    0x7f3e9f70af60 <__GI___strdup>
0x7f3e9f874e88 <strdup@got.plt>:    0x7f3e9f859c10 <baudrate>    0x7f3e9f860550 <_nc_set_no_padding+48>
0x7f3e9f874e98 <_nc_flush_sp@got.plt>:    0x7f3e9f863910 <_nc_visbuf2+16>    0x7f3e9f858b90 <_nc_keep_tic_dir+16>
0x7f3e9f874ea8 <_nc_last_db@got.plt>:    0x7f3e9f85ce00 <tigetnum_sp+256>    0x7f3e9f860e30 <_nc_get_tty_mode_sp+128>
0x7f3e9f874eb8 <_nc_get_tty_mode@got.plt>:    0x7f3e9f724810 <__strstr_sse2_unaligned>    0x7f3e9f863890
0x7f3e9f874ec8 <_nc_visbuf2@got.plt>:    0x7f3e9f85c830 <tgetflag+16>    0x7f3e9f6b3620 <__ctype_b_loc>
0x7f3e9f874ed8 <__ctype_b_loc@got.plt>: 0x7f3e9f859410 <_nc_free_termtype>    0x7f3e9f859d80 <del_curterm+16>
0x7f3e9f874ee8 <_nc_screen_of@got.plt>: 0x7f3e9f78aac0 <___sprintf_chk>    0x7f3e9f87593c <ospeed>
0x7f3e9f874ef8: 0x7f3e9f875538 <_nc_tail>    0x0
0x7f3e9f874f08: 0x55e165168d20 <stdout> 0x7f3e9f8757d0 <SP>
0x7f3e9f874f18: 0x7f3e9f8755c0 <acs_map>    0x7f3e9f875540 <_nc_head>
0x7f3e9f874f28: 0x7f3e9f860750 <delay_output+16>    0x7f3e9f8757d8 <_nc_screen_chain>
0x7f3e9f874f38: 0x7f3e9f875530 <_nc_suppress_warnings> 0x55e165168d48 <UP>
0x7f3e9f874f48: 0x7f3e9f875820 <ttytype>    0x7f3e9f875008 <_nc_user_definable>
0x7f3e9f874f58: 0x7f3e9f8604c0 <tiparm+6208>    0x0
0x7f3e9f874f68: 0x7f3e9f875940 <_nc_tracing>    0x7f3e9f875800 <COLS>
0x7f3e9f874f78: 0x7f3e9f8696a0 <_nc_tinfo_fkeys>    0x7f3e9f87552c <_nc_curr_line>
0x7f3e9f874f88: 0x7f3e9f875930 <_nc_tparm_err> 0x55e165168d50 <PC>
0x7f3e9f874f98: 0x7f3e9f8604e0 <_nc_putchar_sp+16>    0x7f3e9f875804 <LINES>
0x7f3e9f874fa8: 0x7f3e9f8754f0 <TABSIZE>    0x7f3e9f875528 <_nc_curr_col>
0x7f3e9f874fb8: 0x7f3e9f875020 <_nc_prescreen> 0x7f3e9f8757c8 <cur_term>
0x7f3e9f874fc8: 0x0    0x7f3e9f6bd2f0 <__cxa_finalize>
0x7f3e9f874fd8: 0x7f3e9f875320 <_nc_globals>    0x55e165168d60 <BC>
0x7f3e9f874fe8: 0x55e165168d80 <stderr> 0x0
```

Exercise L3 (WinDbg)

Goal: Explore shared libraries and their dependencies.

Memory Analysis Patterns: Module Collection; Coupled Modules; Value References.

Malware Analysis Patterns: Namespace.

1.   Launch WinDbg and load a core dump *core.19649* from the A64 directory:

```
Microsoft (R) Windows Debugger Version 10.0.25324.1001 AMD64
Copyright (c) Microsoft Corporation. All rights reserved.


Loading Dump File [C:\LAPI\A64\core.19649]
64-bit machine not using 64-bit API

************ Path validation summary **************
Response                      Time (ms)     Location
Deferred                                    srv*
Symbol search path is: srv*
Executable search path is:
Generic Unix Version 0 UP Free ARM 64-bit (AArch64)
System Uptime: not available
Process Uptime: not available
.................
*** WARNING: Unable to verify timestamp for libc.so.6
*** WARNING: Unable to verify timestamp for bash
libc_so+0xb6734:
0000ffff`bafa6734 d4000001 svc          #0
```

2.   Set the symbol path and logging to a file in case of lengthy output from some commands:

```
0:000> .sympath+ C:\LAPI\A64
Symbol search path is: srv*;C:\LAPI\A64
Expanded Symbol search path is: cache*;SRV*https://msdl.microsoft.com/download/symbols;c:\
lapi\a64

************ Path validation summary **************
Response                      Time (ms)     Location
Deferred                                    srv*
OK                                          C:\LAPI\A64
*** WARNING: Unable to verify timestamp for libc.so.6
*** WARNING: Unable to verify timestamp for bash

0:000> .reload
.....*** WARNING: Unable to verify timestamp for libc.so.6
..............
*** WARNING: Unable to verify timestamp for bash

************ Symbol Loading Error Summary **************
Module name           Error
bash                  The system cannot find the file specified
libc.so               The system cannot find the file specified
```

121

You can troubleshoot most symbol related issues by turning on symbol loading diagnostics (!sym noisy) and repeating the command that caused symbols to be loaded.
You should also verify that your symbol search path (.sympath) is correct.

```
0:000> .logopen C:\LAPI\A64\L3-WinDbg.log
Opened log file 'C:\LAPI\A64\L3-WinDbg.log'
```

3.   Get the list of loaded shared libraries:

```
0:000> lm
start             end               module name
0000aaaa`bb6a0000 0000aaaa`bb80a000 bash       T (service symbols: ELF Export Symbols)     c:\lapi\a64\bash
0000ffff`babb0000 0000ffff`bac07000 LC_CTYPE T (no symbols)
0000ffff`bac07000 0000ffff`baef0000 locale_archive T (no symbols)
0000ffff`baef0000 0000ffff`bb08e000 libc_so    T (service symbols: ELF Export Symbols)     c:\lapi\a64\libc.so.6
0000ffff`bb0a0000 0000ffff`bb0e0000 libtinfo_so_6 T (service symbols: ELF In Memory Symbols)
0000ffff`bb0e3000 0000ffff`bb0e4000 LC_NUMERIC T (no symbols)
0000ffff`bb0e4000 0000ffff`bb0e5000 LC_TIME   T (no symbols)
0000ffff`bb0e5000 0000ffff`bb0e6000 LC_COLLATE T (no symbols)
0000ffff`bb0e6000 0000ffff`bb0ed000 gconv_modules T (no symbols)
0000ffff`bb0ed000 0000ffff`bb118000 ld_linux_aarch64_so T (service symbols: ELF Export Symbols)     c:\lapi\a64\ld-linux-aarch64.so.1
0000ffff`bb118000 0000ffff`bb119000 LC_MONETARY T (no symbols)
0000ffff`bb11b000 0000ffff`bb11c000 SYS_LC_MESSAGES T (no symbols)
0000ffff`bb11c000 0000ffff`bb11d000 LC_PAPER T (no symbols)
0000ffff`bb11d000 0000ffff`bb11e000 LC_NAME   T (no symbols)
0000ffff`bb11e000 0000ffff`bb11f000 LC_ADDRESS T (no symbols)
0000ffff`bb11f000 0000ffff`bb120000 LC_TELEPHONE T (no symbols)
0000ffff`bb120000 0000ffff`bb121000 LC_MEASUREMENT T (no symbols)
0000ffff`bb121000 0000ffff`bb122000 LC_IDENTIFICATION T (no symbols)
0000ffff`bb126000 0000ffff`bb127000 linux_vdso_so T (service symbols: ELF In Memory Symbols)
```

4.   Get the list of exported functions from all shared libraries having a *free* pattern:

```
0:000> x *!*free*
0000aaaa`bb70d324 bash!unfreeze_jobs_list = <no type information>
0000aaaa`bb798524 bash!rl_free_line_state = <no type information>
0000aaaa`bb779380 bash!rl_free_match_list = <no type information>
0000aaaa`bb6d5280 bash!free_pushed_string_input = <no type information>
0000aaaa`bb79d680 bash!free_history_entry = <no type information>
0000aaaa`bb791de0 bash!rl_free_undo_list = <no type information>
0000aaaa`bb70d310 bash!freeze_jobs_list = <no type information>
0000aaaa`bb7215e0 bash!free_mail_files = <no type information>
0000aaaa`bb7984a0 bash!rl_free_undo_list = <no type information>
0000aaaa`bb740e20 bash!xfree = <no type information>
0000aaaa`bb79cd40 bash!rl_free_history_entry = <no type information>
0000aaaa`bb721d70 bash!free_trap_strings = <no type information>
0000aaaa`bb791b80 bash!rl_free = <no type information>
0000aaaa`bb721fd0 bash!free_buffered_stream = <no type information>
0000aaaa`bb79cde0 bash!rl_free_saved_history_line = <no type information>
0000aaaa`bb779080 bash!rl_free_keymap = <no type information>
0000ffff`bafd72c0 libc_so!pkey_free = <no type information>
0000ffff`bb014860 libc_so!xdr_free = <no type information>
0000ffff`bafab720 libc_so!globfree64 = <no type information>
0000ffff`baf7f5a0 libc_so!obstack_free = <no type information>
0000ffff`bafab720 libc_so!globfree = <no type information>
0000ffff`bafc2100 libc_so!freeaddrinfo = <no type information>
0000ffff`baf7f5a0 libc_so!obstack_free = <no type information>
0000ffff`bb0247a0 libc_so!_libc_freeres = <no type information>
0000ffff`baf23c30 libc_so!_freelocale = <no type information>
0000ffff`bafef3a0 libc_so!if_freenameindex = <no type information>
0000ffff`baf5dff0 libc_so!IO_free_wbackup_area = <no type information>
0000ffff`bafbcc70 libc_so!regfree = <no type information>
0000ffff`baf7dbd4 libc_so!free = <no type information>
0000ffff`baf7dbd4 libc_so!cfree = <no type information>
0000ffff`bb092988 libc_so!_free_hook = <no type information>
0000ffff`baf66e30 libc_so!IO_free_backup_area = <no type information>
```

122

```
0000ffff`baf23c30 libc_so!freelocale = <no type information>
0000ffff`bafc5180 libc_so!wordfree = <no type information>
0000ffff`bafc6d70 libc_so!_sched_cpufree = <no type information>
0000ffff`baf7f6a0 libc_so!_libc_scratch_buffer_dupfree = <no type information>
0000ffff`baff0760 libc_so!freeifaddrs = <no type information>
0000ffff`baf7dbd4 libc_so!_libc_free = <no type information>
0000ffff`bb0b0460 libtinfo_so_6!nc_free_entries = <no type information>
0000ffff`bb0b0450 libtinfo_so_6!nc_free_termtype2 = <no type information>
0000ffff`bb0b0440 libtinfo_so_6!nc_free_termtype = <no type information>
0000ffff`bb0f11e4 ld_linux_aarch64_so!dl_exception_free = <no type information>
```

5.   Find memory locations that reference the function address 0000ffff`baf7dbd4:

```
0:000> s-q 0000aaaa`bb6a0000 L?FFFFFFFFFFFF 0000ffff`baf7dbd4
0000aaaa`bb7ff7e8  0000ffff`baf7dbd4 0000ffff`baf23420
0000ffff`bb08bde8  0000ffff`baf7dbd4 0000ffff`bb0934d8
0000ffff`bb08c028  0000ffff`baf7dbd4 0000ffff`baf16ef0
0000ffff`bb0dee68  0000ffff`baf7dbd4 0000ffff`bafa6ad0
0000ffff`bb128b08  0000ffff`baf7dbd4 0000ffff`baf7d640
```
Note: These can be .GOT locations referenced from .PLT.

6.   Map these locations to module address ranges from the output of the **lm** command:

```
0000aaaa`bb7ff7e8 -      0000aaaa`bb6a0000 0000aaaa`bb80a000   bash
0000ffff`bb08bde8 -      0000ffff`baef0000 0000ffff`bb08e000   libc_so
0000ffff`bb08c028 -      0000ffff`baef0000 0000ffff`bb08e000   libc_so
0000ffff`bb0dee68 -      0000ffff`bb0a0000 0000ffff`bb0e0000   libtinfo_so_6
0000ffff`bb128b08 -      ?
```

Note:: These addresses can also be inspected via the **!address** command:

```
0:000> !address 0000aaaa`bb7ff7e8

Usage:                 Image
Base Address:          0000aaaa`bb7fc000
End Address:           0000aaaa`bb801000
Region Size:           00000000`00005000 (   20.000 kB)
State:                 00001000           MEM_COMMIT
Protect:               00000002           PAGE_READONLY
Type:                  00020000           MEM_PRIVATE
Allocation Base:       0000aaaa`bb7fc000
Allocation Protect:    00000002           PAGE_READONLY
Image Path:            /usr/bin/bash
Module Name:           bash
Loaded Image Name:     bash
Mapped Image Name:
More info:             lmv m bash
More info:             !lmi bash
More info:             ln 0xaaaabb7ff7e8
More info:             !dh 0xaaaabb6a0000


Content source: 1 (target), length: 1818

0:000> !address 0000ffff`bb08bde8

Usage:                 Image
```

123

```
Base Address:              0000ffff`bb088000
End Address:               0000ffff`bb08c000
Region Size:               00000000`00004000 (   16.000 kB)
State:                     00001000          MEM_COMMIT
Protect:                   00000002          PAGE_READONLY
Type:                      00020000          MEM_PRIVATE
Allocation Base:           0000ffff`bb088000
Allocation Protect:        00000002          PAGE_READONLY
Image Path:                /usr/lib/aarch64-linux-gnu/libc.so.6
Module Name:               libc_so
Loaded Image Name:         libc.so.6
Mapped Image Name:
More info:                 lmv m libc_so
More info:                 !lmi libc_so
More info:                 ln 0xffffbb08bde8
More info:                 !dh 0xffffbaef0000


Content source: 1 (target), length: 218


0:000> !address 0000ffff`bb08c028

Usage:                     Image
Base Address:              0000ffff`bb08c000
End Address:               0000ffff`bb08e000
Region Size:               00000000`00002000 (    8.000 kB)
State:                     00001000          MEM_COMMIT
Protect:                   00000004          PAGE_READWRITE
Type:                      00020000          MEM_PRIVATE
Allocation Base:           0000ffff`bb08c000
Allocation Protect:        00000004          PAGE_READWRITE
Image Path:                /usr/lib/aarch64-linux-gnu/libc.so.6
Module Name:               libc_so
Loaded Image Name:         libc.so.6
Mapped Image Name:
More info:                 lmv m libc_so
More info:                 !lmi libc_so
More info:                 ln 0xffffbb08c028
More info:                 !dh 0xffffbaef0000


Content source: 1 (target), length: 1fd8


0:000> !address 0000ffff`bb0dee68

Usage:                     Image
Base Address:              0000ffff`bb0db000
End Address:               0000ffff`bb0df000
Region Size:               00000000`00004000 (   16.000 kB)
State:                     00001000          MEM_COMMIT
Protect:                   00000002          PAGE_READONLY
Type:                      00020000          MEM_PRIVATE
Allocation Base:           0000ffff`bb0db000
Allocation Protect:        00000002          PAGE_READONLY
Image Path:                /usr/lib/aarch64-linux-gnu/libtinfo.so.6.3
Module Name:               libtinfo_so_6
Loaded Image Name:         libtinfo.so.6.3
Mapped Image Name:
More info:                 lmv m libtinfo_so_6
More info:                 !lmi libtinfo_so_6
```

```
More info:               ln 0xffffbb0dee68
More info:               !dh 0xffffbb0a0000


Content source: 1 (target), length: 198


0:000> !address 0000ffff`bb128b08

Usage:                   <unknown>
Base Address:            0000ffff`bb127000
End Address:             0000ffff`bb129000
Region Size:             00000000`00002000 (   8.000 kB)
State:                   00001000          MEM_COMMIT
Protect:                 00000002          PAGE_READONLY
Type:                    00020000          MEM_PRIVATE
Allocation Base:         0000ffff`bb127000
Allocation Protect:      00000002          PAGE_READONLY


Content source: 1 (target), length: 4f8
```

7.   We can see what functions are imported by dumping memory with symbols around such .GOT addresses:

```
0:000> dps 0000ffff`bb0dee68-40 L20
0000ffff`bb0dee28  0000ffff`bafc7030 libc_so!stat
0000ffff`bb0dee30  0000ffff`bafc7c20 libc_so!write
0000ffff`bb0dee38  0000ffff`bafc7d30 libc_so!access
0000ffff`bb0dee40  0000ffff`bafe3cd0 libc_so!_fprintf_chk
0000ffff`bb0dee48  0000ffff`baf80000 libc_so!strcmp
0000ffff`bb0dee50  0000ffff`baf24430 libc_so!_ctype_b_loc
0000ffff`bb0dee58  0000ffff`baf2ed74 libc_so!strtol
0000ffff`bb0dee60  0000ffff`baf59880 libc_so!fread
0000ffff`bb0dee68  0000ffff`baf7dbd4 libc_so!free
0000ffff`bb0dee70  0000ffff`bafa6ad0 libc_so!_nanosleep
0000ffff`bb0dee78  0000ffff`baf7ff40 libc_so!strchr
0000ffff`bb0dee80  0000ffff`baf59db0 libc_so!IO_fwrite
0000ffff`bb0dee88  0000ffff`baf58de0 libc_so!IO_fflush
0000ffff`bb0dee90  0000ffff`baf80180 libc_so!strcpy
0000ffff`bb0dee98  0000ffff`baf80750 libc_so!strncat
0000ffff`bb0deea0  0000ffff`bafccf80 libc_so!tcsetattr
0000ffff`bb0deea8  0000ffff`bafc93b0 libc_so!isatty
0000ffff`bb0deeb0  0000ffff`bafa9060 libc_so!_sysconf
0000ffff`bb0deeb8  0000ffff`bafe57e0 libc_so!_poll_chk
0000ffff`bb0deec0  0000ffff`baf81240 libc_so!strstr
0000ffff`bb0deec8  0000ffff`baf809d0 libc_so!strncpy
0000ffff`bb0deed0  0000ffff`baf17890 libc_so!_errno_location
0000ffff`bb0deed8  0000aaaa`bb7713f0 bash!getenv
0000ffff`bb0deee0  0000ffff`bafcdb80 libc_so!ioctl
0000ffff`bb0deee8  0000ffff`baf21940 libc_so!setlocale
0000ffff`bb0deef0  00000000`0003eaf8
0000ffff`bb0deef8  0000ffff`bb0df63a libtinfo_so_6!ospeed
0000ffff`bb0def00  0000ffff`bb0df850 libtinfo_so_6!nc_tail
0000ffff`bb0def08  00000000`00000000
0000ffff`bb0def10  0000ffff`bb0df640 libtinfo_so_6!SP
0000ffff`bb0def18  0000ffff`bb0df650 libtinfo_so_6!acs_map
0000ffff`bb0def20  0000ffff`baf2d220 libc_so!_cxa_finalize
```

API Usage

- Module usage (static analysis)

  - Hidden Module

- Function usage (dynamic analysis)

```
GDB Commands

(gdb) break function

(gdb) rbreak regex
```

To find out whether a particular module is potentially used, we can look at **Module Collection** or scan in memory for **Hidden Modules**. To find out whether a particular module uses this or that function, we can just look at its .PLT section. It can be done statically, looking at a memory dump, for example. However, if we want to find out where in code a particular function is used, we can't search for its address or its address in the .PLT section because usually relative addressing is used. Here we can find the location in code dynamically, for example, using a breakpoint.

Exercise L4

- **Goal:** Find usage of specific Linux API functions

- **Debugging Implementation Patterns:** Code Breakpoint; Breakpoint Action

- \LAPI-Dumps\Exercise-L4-GDB.pdf

Exercise L4 (GDB)

Goal: Find usage of specific Linux API functions.

Debugging Implementation Patterns: Code Breakpoint; Breakpoint Action.

1.   Launch GDB with */bin/bash* process to debug (you can also choose any other process for experimentation):

```
~/LAPI/x64$ gdb /bin/bash
GNU gdb (Debian 8.2.1-2+b3) 8.2.1
Copyright (C) 2018 Free Software Foundation, Inc.
License GPLv3+: GNU GPL version 3 or later <http://gnu.org/licenses/gpl.html>
This is free software: you are free to change and redistribute it.
There is NO WARRANTY, to the extent permitted by law.
Type "show copying" and "show warranty" for details.
This GDB was configured as "x86_64-linux-gnu".
Type "show configuration" for configuration details.
For bug reporting instructions, please see:
<http://www.gnu.org/software/gdb/bugs/>.
Find the GDB manual and other documentation resources online at:
    <http://www.gnu.org/software/gdb/documentation/>.

For help, type "help".
Type "apropos word" to search for commands related to "word"...
Reading symbols from /bin/bash...(no debugging symbols found)...done.
```

2.   Set logging to a file in case of lengthy output from some commands:

```
(gdb) set logging on L4.log
Copying output to L4.log.
```

3.   Set the breakpoint on the *malloc* function from *libc* (we are specific because there may be other functions with the same name, for example, the loader has its own version of *malloc* to use before *glibc* is loaded):

```
(gdb) break __libc_malloc
Function "__libc_malloc" not defined.
Make breakpoint pending on future shared library load? (y or [n]) y
Breakpoint 1 (__libc_malloc) pending.
```

4.   Run the program, wait for the break-in, and check the backtrace:

```
(gdb) run
Starting program: /bin/bash

Breakpoint 1, __GI___libc_malloc (bytes=5) at malloc.c:3036
3036    malloc.c: No such file or directory.
```

```
(gdb) bt
#0  __GI___libc_malloc (bytes=5) at malloc.c:3036
#1  0x00007ffff7e03ae0 in _nl_normalize_codeset (codeset=codeset@entry=0x7fffffffe7fe "UTF-8",
name_len=name_len@entry=5) at ../intl/l10nflist.c:321
#2  0x00007ffff7dfddb5 in _nl_load_locale_from_archive (category=category@entry=12,
namep=namep@entry=0x7fffffffe180) at loadarchive.c:173
#3  0x00007ffff7dfcd04 in _nl_find_locale (locale_path=0x0, locale_path_len=0,
category=category@entry=12, name=name@entry=0x7fffffffe180)
    at findlocale.c:153
```

```
#4  0x00007ffff7dfc6ef in __GI_setlocale (locale=<optimized out>, category=<optimized out>) at
setlocale.c:340
#5  __GI_setlocale (category=<optimized out>, locale=<optimized out>) at setlocale.c:219
#6  0x00005555555db502 in set_default_locale ()
#7  0x0000555555581ee7 in main ()
```

5.   We continue execution until the next breakpoint is hit:

```
(gdb) continue
Continuing.

Breakpoint 1, __GI___libc_malloc (bytes=5) at malloc.c:3036
3036      in malloc.c

(gdb) bt
#0  __GI___libc_malloc (bytes=5) at malloc.c:3036
#1  0x00007ffff7e03ae0 in _nl_normalize_codeset (codeset=codeset@entry=0x7fffffffe7fe "UTF-8",
name_len=name_len@entry=5) at ../intl/l10nflist.c:321
#2  0x00007ffff7dfddb5 in _nl_load_locale_from_archive (category=category@entry=12,
namep=namep@entry=0x7fffffffe180) at loadarchive.c:173
#3  0x00007ffff7dfcd04 in _nl_find_locale (locale_path=0x0, locale_path_len=0,
category=category@entry=12, name=name@entry=0x7fffffffe180)
    at findlocale.c:153
#4  0x00007ffff7dfc6ef in __GI_setlocale (locale=<optimized out>, category=<optimized out>) at
setlocale.c:340
#5  __GI_setlocale (category=<optimized out>, locale=<optimized out>) at setlocale.c:219
#6  0x00005555555db502 in set_default_locale ()
#7  0x0000555555581ee7 in main ()

(gdb) continue
Continuing.

Breakpoint 1, __GI___libc_malloc (bytes=120) at malloc.c:3036
3036      in malloc.c

(gdb) bt
#0  __GI___libc_malloc (bytes=120) at malloc.c:3036
#1  0x00007ffff7dfdc96 in _nl_load_locale_from_archive (category=category@entry=12,
namep=namep@entry=0x7fffffffe180) at loadarchive.c:460
#2  0x00007ffff7dfcd04 in _nl_find_locale (locale_path=0x0, locale_path_len=0,
category=category@entry=12, name=name@entry=0x7fffffffe180)
    at findlocale.c:153
#3  0x00007ffff7dfc6ef in __GI_setlocale (locale=<optimized out>, category=<optimized out>) at
setlocale.c:340
#4  __GI_setlocale (category=<optimized out>, locale=<optimized out>) at setlocale.c:219
#5  0x00005555555db502 in set_default_locale ()
#6  0x0000555555581ee7 in main ()
```

6.   We can add an action to breakpoints to automate printing backtraces:

```
(gdb) info breakpoints
Num     Type           Disp Enb Address            What
1       breakpoint     keep y   0x00007ffff7e53510 in __GI___libc_malloc at malloc.c:3036
        breakpoint already hit 3 times

(gdb) commands 1
Type commands for breakpoint(s) 1, one per line.
End with a line saying just "end".
>bt
```

```
>continue
>end
```

7.   We now get a backtrace after each breakpoint hit:

```
(gdb) continue
Continuing.

Breakpoint 1, __GI___libc_malloc (bytes=776) at malloc.c:3036
3036     in malloc.c
#0  __GI___libc_malloc (bytes=776) at malloc.c:3036
#1  0x00007ffff7dfd499 in _nl_intern_locale_data (category=category@entry=0,
data=0x7ffff7b005c0, datasize=337024) at loadlocale.c:98
#2  0x00007ffff7dfdd05 in _nl_load_locale_from_archive (category=category@entry=12,
namep=namep@entry=0x7fffffffe180) at loadarchive.c:477
#3  0x00007ffff7dfcd04 in _nl_find_locale (locale_path=0x0, locale_path_len=0,
category=category@entry=12, name=name@entry=0x7fffffffe180)
    at findlocale.c:153
#4  0x00007ffff7dfc6ef in __GI_setlocale (locale=<optimized out>, category=<optimized out>) at
setlocale.c:340
#5  __GI_setlocale (category=<optimized out>, locale=<optimized out>) at setlocale.c:219
#6  0x00005555555db502 in set_default_locale ()
#7  0x0000555555581ee7 in main ()

Breakpoint 1, __GI___libc_malloc (bytes=112) at malloc.c:3036
3036     in malloc.c
#0  __GI___libc_malloc (bytes=112) at malloc.c:3036
#1  0x00007ffff7dfd499 in _nl_intern_locale_data (category=category@entry=1,
data=0x7ffff7dca0d0, datasize=54) at loadlocale.c:98
#2  0x00007ffff7dfdd05 in _nl_load_locale_from_archive (category=category@entry=12,
namep=namep@entry=0x7fffffffe180) at loadarchive.c:477
#3  0x00007ffff7dfcd04 in _nl_find_locale (locale_path=0x0, locale_path_len=0,
category=category@entry=12, name=name@entry=0x7fffffffe180)
    at findlocale.c:153
#4  0x00007ffff7dfc6ef in __GI_setlocale (locale=<optimized out>, category=<optimized out>) at
setlocale.c:340
#5  __GI_setlocale (category=<optimized out>, locale=<optimized out>) at setlocale.c:219
#6  0x00005555555db502 in set_default_locale ()
#7  0x0000555555581ee7 in main ()

Breakpoint 1, __GI___libc_malloc (bytes=1336) at malloc.c:3036
3036     in malloc.c
#0  __GI___libc_malloc (bytes=1336) at malloc.c:3036
#1  0x00007ffff7dfd499 in _nl_intern_locale_data (category=category@entry=2,
data=0x7ffff7dca110, datasize=3284) at loadlocale.c:98
#2  0x00007ffff7dfdd05 in _nl_load_locale_from_archive (category=category@entry=12,
namep=namep@entry=0x7fffffffe180) at loadarchive.c:477
#3  0x00007ffff7dfcd04 in _nl_find_locale (locale_path=0x0, locale_path_len=0,
category=category@entry=12, name=name@entry=0x7fffffffe180)
    at findlocale.c:153
#4  0x00007ffff7dfc6ef in __GI_setlocale (locale=<optimized out>, category=<optimized out>) at
setlocale.c:340
#5  __GI_setlocale (category=<optimized out>, locale=<optimized out>) at setlocale.c:219
#6  0x00005555555db502 in set_default_locale ()
#7  0x0000555555581ee7 in main ()

Breakpoint 1, __GI___libc_malloc (bytes=216) at malloc.c:3036
3036     in malloc.c
#0  __GI___libc_malloc (bytes=216) at malloc.c:3036
```

```
#1  0x00007ffff7dfd499 in _nl_intern_locale_data (category=category@entry=3,
data=0x7ffff7b52a40, datasize=2586242) at loadlocale.c:98
#2  0x00007ffff7dfdd05 in _nl_load_locale_from_archive (category=category@entry=12,
namep=namep@entry=0x7fffffffe180) at loadarchive.c:477
#3  0x00007ffff7dfcd04 in _nl_find_locale (locale_path=0x0, locale_path_len=0,
category=category@entry=12, name=name@entry=0x7fffffffe180)
    at findlocale.c:153
--Type <RET> for more, q to quit, c to continue without paging--q
Quit

(gdb) q
A debugging session is active.

        Inferior 1 [process 23584] will be killed.

Quit anyway? (y or n) y
```

Delayed Dynamic Linking

- ⊙ First call:

```
open@plt:
    0x0000555555581aa0 <+0>:       jmpq    *0xe7aaa(%rip)        # 0x555555669550 <open@got.plt>
    0x0000555555581aa6 <+6>:       pushq   $0xa7
    0x0000555555581aab <+11>:      jmpq    0x555555581020
```

DL

- ⊙ Subsequent calls:

LIBC

```
open@plt:
    0x0000555555581aa0 <+0>:       jmpq    *0xe7aaa(%rip)        # 0x555555669550 <open@got.plt>
    0x0000555555581aa6 <+6>:       pushq   $0xa7
    0x0000555555581aab <+11>:      jmpq    0x555555581020
```

If certain Linux API function categories may not be used (for example, lazy evaluation) at all during the program execution, there is no need to resolve .PLT references to the loaded shared library. You may have noticed it in the previous exercises with some LIBC functions having *@plt+offset* in the .GOT.PLT section as a destination. The second jump will go to a dynamic linker and replace the value to point to the real address from the shared library. We see that in the next live debugging exercise.

Exercise L5

- **Goal:** Explore the delayed dynamic linking

- **Debugging Implementation Patterns:** Code Breakpoint; Break-in

- **ADDR Patterns:** Call Path

- \LAPI-Dumps\Exercise-L5-GDB.pdf

Exercise L5 (GDB)

Goal: Explore the delayed dynamic linking.

Debugging Implementation Patterns: Code Breakpoint; Break-in.

ADDR Patterns: Call Path.

1.   Launch GDB with */bin/bash* process to debug (you can also choose any other process for experimentation):

```
~/LAPI/x64$ gdb /bin/bash
GNU gdb (Debian 8.2.1-2+b3) 8.2.1
Copyright (C) 2018 Free Software Foundation, Inc.
License GPLv3+: GNU GPL version 3 or later <http://gnu.org/licenses/gpl.html>
This is free software: you are free to change and redistribute it.
There is NO WARRANTY, to the extent permitted by law.
Type "show copying" and "show warranty" for details.
This GDB was configured as "x86_64-linux-gnu".
Type "show configuration" for configuration details.
For bug reporting instructions, please see:
<http://www.gnu.org/software/gdb/bugs/>.
Find the GDB manual and other documentation resources online at:
    <http://www.gnu.org/software/gdb/documentation/>.

For help, type "help".
Type "apropos word" to search for commands related to "word"...
Reading symbols from /bin/bash...(no debugging symbols found)...done.
```

2.   Set logging to a file in case of lengthy output from some commands:

```
(gdb) set logging on L5.log
Copying output to L5.log.
```

3.   Set the breakpoint on the main function and run the program to allow shared libraries to be loaded:

```
(gdb) break main
Breakpoint 1 at 0x2de30
```

```
(gdb) r
Starting program: /bin/bash

Breakpoint 1, 0x0000555555581e30 in main ()
```

4.   Disassemble the *shell_execve* function and follow the call path of *open@plt*:

```
(gdb) disassemble shell_execve
Dump of assembler code for function shell_execve:
   0x00005555555987a0 <+0>:     push   %r15
   0x00005555555987a2 <+2>:     push   %r14
   0x00005555555987a4 <+4>:     mov    %rdx,%r14
   0x00005555555987a7 <+7>:     push   %r13
   0x00005555555987a9 <+9>:     mov    %rsi,%r13
   0x00005555555987ac <+12>:    push   %r12
   0x00005555555987ae <+14>:    push   %rbp
   0x00005555555987af <+15>:    push   %rbx
```

134

```
   0x00005555555987b0 <+16>:    mov     %rdi,%rbx
   0x00005555555987b3 <+19>:    sub     $0xa8,%rsp
   0x00005555555987ba <+26>:    mov     %fs:0x28,%rax
   0x00005555555987c3 <+35>:    mov     %rax,0x98(%rsp)
   0x00005555555987cb <+43>:    xor     %eax,%eax
   0x00005555555987cd <+45>:    callq   0x5555555815e0 <execve@plt>
   0x00005555555987d2 <+50>:    callq   0x555555581120 <__errno_location@plt>
   0x00005555555987d7 <+55>:    mov     %rax,%r12
   0x00005555555987da <+58>:    mov     (%rax),%ebp
   0x00005555555987dc <+60>:    mov     0xdc2a6(%rip),%eax        # 0x555555674a88
<terminating_signal>
   0x00005555555987e2 <+66>:    test    %eax,%eax
   0x00005555555987e4 <+68>:    jne     0x555555598880 <shell_execve+224>
   0x00005555555987ea <+74>:    cmp     $0x8,%ebp
   0x00005555555987ed <+77>:    je      0x555555598894 <shell_execve+244>
   0x00005555555987f3 <+83>:    xor     %eax,%eax
   0x00005555555987f5 <+85>:    cmp     $0x2,%ebp
   0x00005555555987f8 <+88>:    mov     %rbx,%rdi
   0x00005555555987fb <+91>:    sete    %al
   0x00005555555987fe <+94>:    add     $0x7e,%eax
   0x0000555555598801 <+97>:    mov     %eax,0xde321(%rip)        # 0x555555676b28
<last_command_exit_value>
   0x0000555555598807 <+103>:   callq   0x5555555921a0 <file_isdir>
   0x000055555559880c <+108>:   test    %eax,%eax
   0x000055555559880e <+110>:   jne     0x555555598a58 <shell_execve+696>
   0x0000555555598814 <+116>:   mov     %rbx,%rdi
   0x0000555555598817 <+119>:   callq   0x5555555dc970 <executable_file>
   0x000055555559881c <+124>:   test    %eax,%eax
   0x000055555559881e <+126>:   je      0x555555598841 <shell_execve+161>
   0x0000555555598820 <+128>:   cmp     $0x7,%ebp
   0x0000555555598823 <+131>:   je      0x555555598841 <shell_execve+161>
   0x0000555555598825 <+133>:   cmp     $0xc,%ebp
   0x0000555555598828 <+136>:   je      0x555555598841 <shell_execve+161>
   0x000055555559882a <+138>:   xor     %esi,%esi
   0x000055555559882c <+140>:   mov     %rbx,%rdi
   0x000055555559882f <+143>:   xor     %eax,%eax
   0x0000555555598831 <+145>:   callq   0x555555581aa0 <open@plt>
   0x0000555555598836 <+150>:   mov     %eax,%r13d
--Type <RET> for more, q to quit, c to continue without paging--q
Quit


(gdb) disassemble 0x555555581aa0
Dump of assembler code for function open@plt:
   0x0000555555581aa0 <+0>:     jmpq    *0xe7aaa(%rip)          # 0x555555669550 <open@got.plt>
   0x0000555555581aa6 <+6>:     pushq   $0xa7
   0x0000555555581aab <+11>:    jmpq    0x555555581020
End of assembler dump.


(gdb) x/a 0x555555669550
0x555555669550 <open@got.plt>:  0x555555581aa6 <open@plt+6>


(gdb) disassemble 0x555555581020
No function contains specified address.


(gdb) x/3i 0x555555581020
   0x555555581020:      pushq   0xe7fe2(%rip)        # 0x555555669008
   0x555555581026:      jmpq    *0xe7fe4(%rip)       # 0x555555669010
   0x55555558102c:      nopl    0x0(%rax)
```

```
(gdb) x/a 0x555555669010
0x555555669010: 0x7ffff7fea510 <_dl_runtime_resolve_xsavec>
```

Note: We see that the call path goes back to .PLT section and then jumps to a runtime resolution function from the *dl* library. This means that the *open* function was not yet called from the code. You can recognize not yet-called functions by the pattern in .GOT/.GOT.PLT sections: ...*@plt+6*.

5.   Continue running the program, then break-in by ^C:

```
(gdb) continue
Continuing.
[Detaching after fork from child process 23845]
coredump@DESKTOP-IS6V2L0:~/LAPI/x64$ ^C
Program received signal SIGINT, Interrupt.
0x00007ffff7ebfc89 in __pselect (nfds=1, readfds=0x7fffffffd130, writefds=0x0, exceptfds=0x0,
timeout=<optimized out>,
    sigmask=0x55555567b840 <_rl_orig_sigset>) at ../sysdeps/unix/sysv/linux/pselect.c:69
69      ../sysdeps/unix/sysv/linux/pselect.c: No such file or directory.
(gdb)
```

6.   Now we disassemble the .GOT.PLT section data again and see the real function address:

```
(gdb) disassemble 0x555555581aa0
Dump of assembler code for function open@plt:
   0x0000555555581aa0 <+0>:     jmpq   *0xe7aaa(%rip)        # 0x555555669550 <open@got.plt>
   0x0000555555581aa6 <+6>:     pushq  $0xa7
   0x0000555555581aab <+11>:    jmpq   0x555555581020
End of assembler dump.
```

```
(gdb) x/a 0x555555669550
0x555555669550 <open@got.plt>:   0x7ffff7eb9010 <__libc_open64>
```

Note: Now we see that sometimes in the past, when the *open* function was called for the first time, the dynamic linker resolved the indirect address to the actual address in *libc*. You can verify this by setting the breakpoint on the 0x555555581aa0 address and restarting the program:

```
(gdb) break *0x555555581aa0
Breakpoint 2 at 0x555555581aa0
```

```
(gdb) run
The program being debugged has been started already.
Start it from the beginning? (y or n) y
Starting program: /bin/bash

Breakpoint 1, 0x0000555555581e30 in main ()
```

```
(gdb) x/a 0x555555669550
0x555555669550 <open@got.plt>:   0x555555581aa6 <open@plt+6>
```

```
(gdb) continue
Continuing.

Breakpoint 2, 0x0000555555581aa0 in open@plt ()
```

```
(gdb) x/a 0x555555669550
0x555555669550 <open@got.plt>:   0x555555581aa6 <open@plt+6>
```

```
(gdb) continue
Continuing.

Breakpoint 2, 0x0000555555581aa0 in open@plt ()

(gdb) x/a 0x555555669550
0x555555669550 <open@got.plt>:    0x7ffff7eb9010 <__libc_open64>
```

API Sequences (Prescriptive)

- open, …, close

- malloc, …, free

- pthread_create, …, pthread_join

- socket, …, bind, …, listen, …, accept

- wl_surface_damage, …, wl_surface_attach, …, wl_surface_commit

By prescriptive API sequences, we mean how to use API functions and in what sequence. It is like a traditional approach to language grammar teaching how to use a language. The slide shows some iconic prescriptive sequences we talk about when we discuss appropriate API classes.

API Sequences (Descriptive)

◉ Horizontal

- Code disassembly
- Traces and logs (Thread of Activity analysis pattern)

◉ Vertical

- Stack trace
- Traces and logs (Fiber Bundle analysis pattern)

Descriptive sequences are sequences that we actually see in memory dumps, traces, and logs, and also during live debugging. They may deviate from prescriptive sequences, which should trigger an investigation if detected. We can discover and analyze horizontal API sequences by disassembling code or looking at trace and log messages corresponding to particular threads, and vertical API sequences by looking at stack traces from problem threads during postmortem debugging, API usage during live debugging sessions, or looking at stack traces associated with particular trace and log messages, the so-called **Fiber Bundle** trace and log analysis pattern.

Thread of Activity
https://www.dumpanalysis.org/blog/index.php/2009/08/03/trace-analysis-patterns-part-7/

Fiber Bundle
https://www.dumpanalysis.org/blog/index.php/2012/09/26/trace-analysis-patterns-part-52/

API Layers

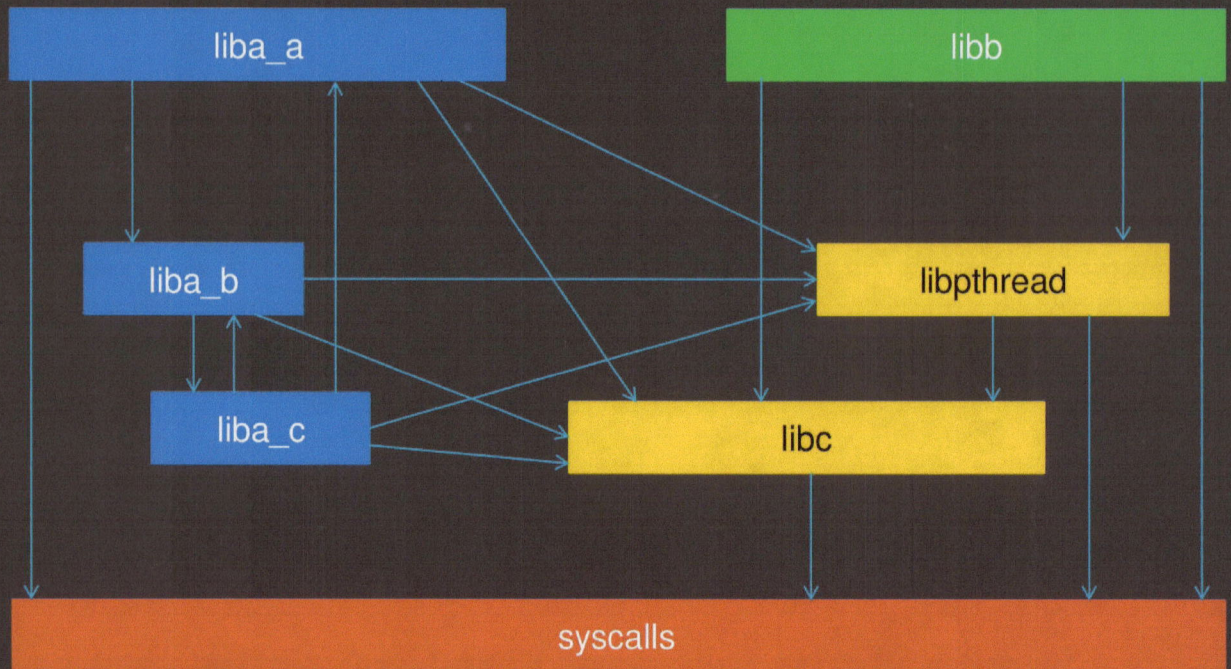

In complex applications and services, there can be several API layers. The top layers may reference the lower layers and also the lowest ones for basic functions. The *libc* layer is usually considered the lowest API level that contains function wrappers for system calls (syscalls). However, any library and executable itself can call syscalls directly as well.

API Internals

- Memory analysis patterns:

  - Hooked Functions (User Space)
  - Module patterns

    - Hooked Modules

- Malware analysis patterns:

  - Patched Code

GDB Commands

```
(gdb) disassemble function

(gdb) x/<n>i address
```

WinDbg Commands

```
0:000> !for_each_module

0:000> u fname

0:000> uf /c fname
```

Linux API functions can be patched by value-adding code and malware. Also, the caller of API may be affected by the order of loaded shared libraries. The next exercise covers ADDR patterns such as **Function Skeleton** and **Call Path**.

API and System Calls

◉ <u>Syscall numbers and arguments</u>

◉ API that does not require syscalls

- strlen
- malloc
- fread

◉ API that requires kernel services

- getpid
- mmap
- read

Sometimes, we want to investigate the code of an API function to know what other API functions it uses and whether it is translated to a system call. For example, some API functions do not require kernel services, such as the current process and thread ids. Other functions may cache the result from the kernel or the different execution paths without kernel services.

Syscall numbers and arguments
https://syscall.sh/

Exercise L6

- **Goal:** Explore API layers and internals of specific API functions; check whether the selected API functions use system calls

- **ADDR Patterns:** Function Skeleton; Call Path

- \LAPI-Dumps\Exercise-L6-GDB.pdf

- \LAPI-Dumps\Exercise-L6-WinDbg.pdf

Exercise L6 (GDB)

Goal: Explore API layers and internals of specific API functions; check whether the selected API functions use system calls.

ADDR Patterns: Function Skeleton; Call Path.

1.   Load a core dump *core.9* and *bash* executable from the x64 directory:

```
~/LAPI/x64$ gdb -c core.9 -se bash
GNU gdb (Debian 8.2.1-2+b3) 8.2.1
Copyright (C) 2018 Free Software Foundation, Inc.
License GPLv3+: GNU GPL version 3 or later <http://gnu.org/licenses/gpl.html>
This is free software: you are free to change and redistribute it.
There is NO WARRANTY, to the extent permitted by law.
Type "show copying" and "show warranty" for details.
This GDB was configured as "x86_64-linux-gnu".
Type "show configuration" for configuration details.
For bug reporting instructions, please see:
<http://www.gnu.org/software/gdb/bugs/>.
Find the GDB manual and other documentation resources online at:
    <http://www.gnu.org/software/gdb/documentation/>.

For help, type "help".
Type "apropos word" to search for commands related to "word"...
Reading symbols from bash...(no debugging symbols found)...done.

warning: core file may not match specified executable file.
[New LWP 9]
Core was generated by `-bash'.
#0  0x00007f3e9f7492d7 in __GI___waitpid (pid=-1, stat_loc=0x7ffcbd661ad0, options=10) at
../sysdeps/unix/sysv/linux/waitpid.c:30
30       ../sysdeps/unix/sysv/linux/waitpid.c: No such file or directory.
```

2.   Set logging to a file in case of lengthy output from some commands:

```
(gdb) set logging on L6.log
Copying output to L6.log.
```

3.   Let's follow the call path of the *_nc_is_dir_path* function from the *libtinfo.so* library:

```
(gdb) disassemble _nc_is_dir_path
Dump of assembler code for function _nc_is_dir_path:
   0x00007f3e9f856b50 <+0>:     sub     $0xa8,%rsp
   0x00007f3e9f856b57 <+7>:     mov     %rdi,%rsi
   0x00007f3e9f856b5a <+10>:    mov     $0x1,%edi
   0x00007f3e9f856b5f <+15>:    mov     %fs:0x28,%rax
   0x00007f3e9f856b68 <+24>:    mov     %rax,0x98(%rsp)
   0x00007f3e9f856b70 <+32>:    xor     %eax,%eax
   0x00007f3e9f856b72 <+34>:    mov     %rsp,%rdx
   0x00007f3e9f856b75 <+37>:    callq   0x7f3e9f856550 <__xstat@plt>
   0x00007f3e9f856b7a <+42>:    xor     %edx,%edx
   0x00007f3e9f856b7c <+44>:    test    %eax,%eax
   0x00007f3e9f856b7e <+46>:    jne     0x7f3e9f856b91 <_nc_is_dir_path+65>
   0x00007f3e9f856b80 <+48>:    mov     0x18(%rsp),%eax
   0x00007f3e9f856b84 <+52>:    and     $0xf000,%eax
```

144

```
   0x00007f3e9f856b89 <+57>:    cmp     $0x4000,%eax
   0x00007f3e9f856b8e <+62>:    sete    %dl
   0x00007f3e9f856b91 <+65>:    mov     0x98(%rsp),%rcx
   0x00007f3e9f856b99 <+73>:    xor     %fs:0x28,%rcx
   0x00007f3e9f856ba2 <+82>:    mov     %edx,%eax
   0x00007f3e9f856ba4 <+84>:    jne     0x7f3e9f856bae <_nc_is_dir_path+94>
   0x00007f3e9f856ba6 <+86>:    add     $0xa8,%rsp
   0x00007f3e9f856bad <+93>:    retq
   0x00007f3e9f856bae <+94>:    callq   0x7f3e9f8562a0 <__stack_chk_fail@plt>
End of assembler dump.


(gdb) disassemble 0x7f3e9f856550
Dump of assembler code for function __xstat@plt:
   0x00007f3e9f856550 <+0>:     jmpq    *0x1e7aa(%rip)          # 0x7f3e9f874d00
<__xstat@got.plt>
   0x00007f3e9f856556 <+6>:     pushq   $0x52
   0x00007f3e9f85655b <+11>:    jmpq    0x7f3e9f856020
End of assembler dump.


(gdb) x/a 0x7f3e9f874d00
0x7f3e9f874d00 <__xstat@got.plt>:       0x7f3e9f76c940 <__GI___xstat>


(gdb) disassemble 0x7f3e9f76c940
Dump of assembler code for function __GI___xstat:
   0x00007f3e9f76c940 <+0>:     mov     %rsi,%rax
   0x00007f3e9f76c943 <+3>:     cmp     $0x1,%edi
   0x00007f3e9f76c946 <+6>:     ja      0x7f3e9f76c978 <__GI___xstat+56>
   0x00007f3e9f76c948 <+8>:     mov     %rax,%rdi
   0x00007f3e9f76c94b <+11>:    mov     %rdx,%rsi
   0x00007f3e9f76c94e <+14>:    mov     $0x4,%eax
   0x00007f3e9f76c953 <+19>:    syscall
   0x00007f3e9f76c955 <+21>:    cmp     $0xfffffffffffff000,%rax
   0x00007f3e9f76c95b <+27>:    ja      0x7f3e9f76c960 <__GI___xstat+32>
   0x00007f3e9f76c95d <+29>:    retq
   0x00007f3e9f76c95e <+30>:    xchg    %ax,%ax
   0x00007f3e9f76c960 <+32>:    mov     0xd0509(%rip),%rdx      # 0x7f3e9f83ce70
   0x00007f3e9f76c967 <+39>:    neg     %eax
   0x00007f3e9f76c969 <+41>:    mov     %eax,%fs:(%rdx)
   0x00007f3e9f76c96c <+44>:    mov     $0xffffffff,%eax
   0x00007f3e9f76c971 <+49>:    retq
   0x00007f3e9f76c972 <+50>:    nopw    0x0(%rax,%rax,1)
   0x00007f3e9f76c978 <+56>:    mov     0xd04f1(%rip),%rax      # 0x7f3e9f83ce70
   0x00007f3e9f76c97f <+63>:    movl    $0x16,%fs:(%rax)
   0x00007f3e9f76c986 <+70>:    mov     $0xffffffff,%eax
   0x00007f3e9f76c98b <+75>:    retq
End of assembler dump.
```

Note: We see that the _nc_is_dir_path function from the *libtinfo.so* library calls the _xstat function from the *libc.so* library, and the latter uses the syscall number 4 (stat).

4.   Load a core dump *core.19649* and *bash* executable from the A64 directory:

~/LAPI/x64$ **cd ../A64**

~/LAPI/A64$ **gdb-multiarch -c core.19649 -se bash**
GNU gdb (Debian 8.2.1-2+b3) 8.2.1
Copyright (C) 2018 Free Software Foundation, Inc.
License GPLv3+: GNU GPL version 3 or later <http://gnu.org/licenses/gpl.html>

145

This is free software: you are free to change and redistribute it.
There is NO WARRANTY, to the extent permitted by law.
Type "show copying" and "show warranty" for details.
This GDB was configured as "x86_64-linux-gnu".
Type "show configuration" for configuration details.
For bug reporting instructions, please see:
<http://www.gnu.org/software/gdb/bugs/>.
Find the GDB manual and other documentation resources online at:
    <http://www.gnu.org/software/gdb/documentation/>.

For help, type "help".
Type "apropos word" to search for commands related to "word"...
Reading symbols from bash...(no debugging symbols found)...done.

warning: core file may not match specified executable file.
[New LWP 19649]

warning: Could not load shared library symbols for 3 libraries, e.g.
/lib/aarch64-linux-gnu/libtinfo.so.6.
Use the "info sharedlibrary" command to see the complete listing.
Do you need "set solib-search-path" or "set sysroot"?
Core was generated by `-bash'.
#0 0x0000ffffbafa6734 in ?? ()

(gdb) **set solib-search-path .**
Reading symbols from /home/coredump/LAPI/A64/libtinfo.so.6...(no debugging symbols found)...done.
Reading symbols from /home/coredump/LAPI/A64/libc.so.6...(no debugging symbols found)...done.
Reading symbols from /home/coredump/LAPI/A64/ld-linux-aarch64.so.1...(no debugging symbols found)...done.

5.   Set logging to a file in case of lengthy output from some commands:

(gdb) **set logging on** L6.log
Copying output to L6.log.

6.   Disassemble the *getpid* and *getppid* functions and find their syscall numbers:

(gdb) **disassemble** getpid
Dump of assembler code for function getpid:
   0x0000ffffbafa8000 <+0>:   nop
   0x0000ffffbafa8004 <+4>:   mov   x8, #0xac   // #172
   0x0000ffffbafa8008 <+8>:   svc   #0x0
   0x0000ffffbafa800c <+12>:   ret
End of assembler dump.

(gdb) **disassemble** getppid
Dump of assembler code for function getppid:
   0x0000ffffbafa8040 <+0>:   nop
   0x0000ffffbafa8044 <+4>:   mov   x8, #0xad   // #173
   0x0000ffffbafa8048 <+8>:   svc   #0x0
   0x0000ffffbafa804c <+12>:   ret
End of assembler dump.

Exercise L6 (WinDbg)

Goal: Explore API layers and internals of specific API functions; check whether the selected API functions use system calls.

ADDR Patterns: Function Skeleton; Call Path.

1.   Launch WinDbg and load a core dump *core.19649* from the A64 directory:

```
Microsoft (R) Windows Debugger Version 10.0.25324.1001 AMD64
Copyright (c) Microsoft Corporation. All rights reserved.


Loading Dump File [C:\LAPI\A64\core.19649]
64-bit machine not using 64-bit API

************* Path validation summary **************
Response                        Time (ms)     Location
Deferred                                      srv*
Symbol search path is: srv*
Executable search path is:
Generic Unix Version 0 UP Free ARM 64-bit (AArch64)
System Uptime: not available
Process Uptime: not available
.................
*** WARNING: Unable to verify timestamp for libc.so.6
*** WARNING: Unable to verify timestamp for bash
libc_so+0xb6734:
0000ffff`bafa6734 d4000001 svc            #0
```

2.   Set the symbol path and logging to a file in case of lengthy output from some commands:

```
0:000> .sympath+ C:\LAPI\A64
Symbol search path is: srv*;C:\LAPI\A64
Expanded Symbol search path is: cache*;SRV*https://msdl.microsoft.com/download/symbols;c:\
lapi\a64

************* Path validation summary **************
Response                        Time (ms)     Location
Deferred                                      srv*
OK                                            C:\LAPI\A64
*** WARNING: Unable to verify timestamp for libc.so.6
*** WARNING: Unable to verify timestamp for bash

0:000> .reload
.....*** WARNING: Unable to verify timestamp for libc.so.6
.............
*** WARNING: Unable to verify timestamp for bash

************* Symbol Loading Error Summary **************
Module name        Error
bash               The system cannot find the file specified
libc.so            The system cannot find the file specified
```

You can troubleshoot most symbol related issues by turning on symbol loading diagnostics (!sym noisy) and repeating the command that caused symbols to be loaded.

You should also verify that your symbol search path (.sympath) is correct.

```
0:000> .logopen C:\LAPI\A64\L6-WinDbg.log
Opened log file 'C:\LAPI\A64\L6-WinDbg.log'
```

3.   Let's see the function skeleton of the *nc_is_dir_path* function from the *libtinfo_so_6* library and follow the call path of the first branch and link:

```
0:000> uf /c libtinfo_so_6!nc_is_dir_path
libtinfo_so_6!nc_is_dir_path (0000ffff`bb0ae9c0)
  libtinfo_so_6!nc_is_dir_path+0x20 (0000ffff`bb0ae9e0):
    call to libtinfo_so_6!+0x230 (0000ffff`bb0ad6d0)
  libtinfo_so_6!nc_is_dir_path+0x64 (0000ffff`bb0aea24):
    call to libtinfo_so_6!+0x200 (0000ffff`bb0ad6a0)
  libtinfo_so_6!nc_is_file_path+0x20 (0000ffff`bb0aea50):
    call to libtinfo_so_6!+0x230 (0000ffff`bb0ad6d0)
  libtinfo_so_6!nc_is_file_path+0x64 (0000ffff`bb0aea94):
    call to libtinfo_so_6!+0x200 (0000ffff`bb0ad6a0)
  libtinfo_so_6!nc_add_to_try+0x8c (0000ffff`bb0aeb2c):
    call to libtinfo_so_6!+0x1d0 (0000ffff`bb0ad670)
  libtinfo_so_6!nc_add_to_try+0xe8 (0000ffff`bb0aeb88):
    call to libtinfo_so_6!+0x1d0 (0000ffff`bb0ad670)
  libtinfo_so_6!nc_add_to_try+0x130 (0000ffff`bb0aebd0):
    call to libtinfo_so_6!+0x1d0 (0000ffff`bb0ad670)
  libtinfo_so_6!nc_add_to_try+0x160 (0000ffff`bb0aec00):
    call to libtinfo_so_6!+0x2b0 (0000ffff`bb0ad750)
```

```
0:000> u 0000ffff`bb0ad6d0
libtinfo_so_6!+0x230:
0000ffff`bb0ad6d0 b0000190 adrp    xip0,libtinfo_so_6!strfnames+0x200 (0000ffff`bb0de000)
0000ffff`bb0ad6d4 f9471611 ldr     xip1,[xip0,#0xE28]
0000ffff`bb0ad6d8 9138a210 add     xip0,xip0,#0xE28
0000ffff`bb0ad6dc d61f0220 br      xip1
0000ffff`bb0ad6e0 b0000190 adrp    xip0,libtinfo_so_6!strfnames+0x200 (0000ffff`bb0de000)
0000ffff`bb0ad6e4 f9471a11 ldr     xip1,[xip0,#0xE30]
0000ffff`bb0ad6e8 9138c210 add     xip0,xip0,#0xE30
0000ffff`bb0ad6ec d61f0220 br      xip1
```

```
0:000> dps 0000ffff`bb0de000+0xE28 L1
0000ffff`bb0dee28  0000ffff`bafc7030 libc_so!stat
```

```
0:000> uf libc_so!stat
libc_so!stat:
0000ffff`bafc7030 aa0103e2 mov     x2,x1
0000ffff`bafc7034 52800003 mov     w3,#0
0000ffff`bafc7038 aa0003e1 mov     x1,x0
0000ffff`bafc703c 12800c60 mov     w0,#-0x64
0000ffff`bafc7040 1400001c b       libc_so!fstatat (0000ffff`bafc70b0)  Branch

libc_so!fstatat:
0000ffff`bafc70b0 93407c00 sxtw    x0,w0
0000ffff`bafc70b4 93407c63 sxtw    x3,w3
0000ffff`bafc70b8 d28009e8 mov     x8,#0x4F
0000ffff`bafc70bc d4000001 svc     #0
0000ffff`bafc70c0 3140041f cmn     w0,#1,lsl #0xC
0000ffff`bafc70c4 54000068 bhi     libc_so!fstatat+0x20 (0000ffff`bafc70d0)  Branch

libc_so!fstatat+0x18:
0000ffff`bafc70c8 52800000 mov     w0,#0
```

```
0000ffff`bafc70cc d65f03c0 ret

libc_so!fstatat+0x20:
0000ffff`bafc70d0 90000622 adrp        x2,libc_so!sys_siglist+0x1a8 (0000ffff`bb08b000)
0000ffff`bafc70d4 f946e442 ldr         x2,[x2,#0xDC8]
0000ffff`bafc70d8 d53bd043 mrs         x3,TPIDR_EL0
0000ffff`bafc70dc 4b0003e1 neg         w1,w0
0000ffff`bafc70e0 12800000 mov         w0,#-1
0000ffff`bafc70e4 b8226861 str         w1,[x3,x2]
0000ffff`bafc70e8 d65f03c0 ret
```

Note: We see that the *nc_is_dir_path* function from the *libtinfo.so* library calls the *stat* function from the *libc.so* library, and the latter uses the syscall number 0x4F (stat).

4.   Disassemble the *getpid* and *getppid* functions and find their syscall numbers:

```
0:000> uf getpid
libc_so!getpid:
0000ffff`bafa8000 d503201f nop
0000ffff`bafa8004 d2801588 mov         x8,#0xAC
0000ffff`bafa8008 d4000001 svc         #0
0000ffff`bafa800c d65f03c0 ret

0:000> uf getppid
libc_so!getppid:
0000ffff`bafa8040 d503201f nop
0000ffff`bafa8044 d28015a8 mov         x8,#0xAD
0000ffff`bafa8048 d4000001 svc         #0
0000ffff`bafa804c d65f03c0 ret
```

API Name Patterns

- create/open/delete/close

- pthread

- display/surface/window

- alloc/free

- read/write

GDB Commands
`(gdb) info functions pattern`

WinDbg Commands
`0:000> x module!fpattern`

One way to learn about various available API functions is to search for name patterns, such as all functions that have some substring in their symbol name.

API Namespaces

- Functions required to accomplish a particular task

  - Example: network communication

API Namespace is a group or several groups of functions required to accomplish a particular task. A namespace group usually belongs to one particular shared library. As we mentioned before, malware analysis patterns include the **Namespace** analysis pattern that views groups of imported functions from several shared libraries as potentially serving some particular malicious (but maybe just some utilitarian) need, for example, network communication or a screen capture that needs to be saved somewhere.

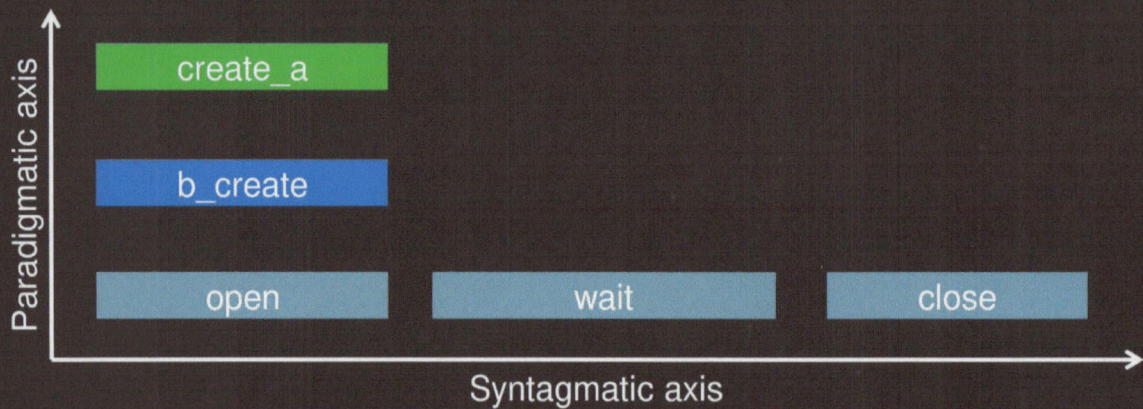

API Syntagms/Paradigms

- Syntagms / syntagmatic analysis

- Paradigms / paradigmatic analysis

© 2023 Software Diagnostics Services

API Namespace is a group or several groups of functions required to accomplish a particular task. A namespace group usually belongs to one particular shared library. As we mentioned before, malware analysis patterns include the **Namespace** analysis pattern that views groups of imported functions from several shared libraries as potentially serving some particular malicious (but maybe just some utilitarian) need, for example, network communication or a screen capture that needs to be saved somewhere.

Syntagms
https://en.wikipedia.org/wiki/Syntagma_(linguistics)

Syntagmatic analysis
https://en.wikipedia.org/wiki/Syntagmatic_analysis

Paradigmatic analysis
https://en.wikipedia.org/wiki/Paradigmatic_analysis

Marked API

- Marked Message trace and log analysis pattern

- Points to the presence or absence of activity

- Example:

  - execve [-]
  - socket [+]
  - connect [+]
  - create [-]

> **GDB Commands**
> ```
> (gdb) info functions @plt
> ```

We call by marked API certain functions imported or present in call paths that point to the presence of activity. If they are not imported and not used, they point to the absence of activity. Such API functions may be compiled into a checklist. This API aspect is borrowed from **Marked Message** trace and log analysis pattern where marked messages may point to some domain of software activity related to functional requirements and help in troubleshooting and debugging, and some unmarked messages may directly say about the absences of activity.

Marked Message
https://www.dumpanalysis.org/blog/index.php/2012/01/02/trace-analysis-patterns-part-45/

ADDR Patterns

- From **A**ccelerated **D**isassembly **D**econstruction **R**eversing

- List of pattern names

- Pattern descriptions

The name ADDR (sounds like an address) comes from the Accelerated Disassembly Deconstruction Reversing abbreviation from the similar sounding training course. The slide provides the link to their names and the link to their descriptions. We used some ADDR patterns, such as **Call Path** and **Function Skeleton**, in some of our exercises.

List of pattern names
https://www.dumpanalysis.org/addr-patterns

Pattern descriptions
https://www.patterndiagnostics.com/Training/Accelerated-Linux-Disassembly-Reconstruction-Reversing-Second-Edition-Slides.pdf

DebugWare Patterns

- Patterns for troubleshooting and debugging tools

- API Query

  Periodic or asynchronous query of the same set of APIs and logging of their input and output data.

DebugWare patterns are patterns for designing and implementing troubleshooting and debugging tools. One of the first patterns was named **API Query**, where some API functions are used periodically to query OS data and log the output.

API Query
https://www.dumpanalysis.org/blog/index.php/2008/07/19/debugware-patterns-part-1/

Patterns vs. Analysis Patterns

Diagnostic Pattern: a common recurrent identifiable problem together with a set of recommendations and possible solutions to apply in a specific context.

Diagnostic Problem: a set of indicators (symptoms, signs) describing a problem.

Diagnostic Analysis Pattern: a common recurrent analysis technique and method of diagnostic pattern identification in a specific context.

Diagnostics Pattern Language: common names of diagnostic and diagnostic analysis patterns. The same language for any operating system: Windows, macOS, Linux, ...

We have now come to the association of Linux API with various diagnostic analysis patterns. Here, I'd like to stress the informal difference between patterns and analysis patterns. The distinction is highlighted on this slide.

Memory Analysis Patterns

- ⊙ **User space**

  - Process memory dumps

- ⊙ **Function analysis patterns**

  - Stack Trace Collection
  - Well-Tested Function
  - False Function Parameters
  - String Parameter
  - Small Value / Design Value

  - Stack Trace
  - Execution Residue
  - Hidden Parameter
  - Parameter Flow
  - Data Correlation

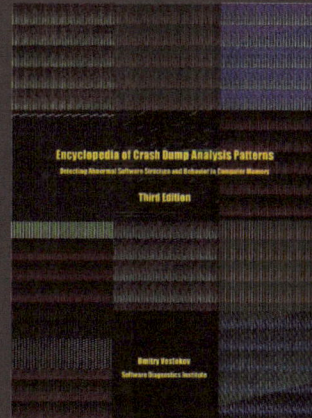

We pay attention to Linux API function calls when we do diagnostics and postmortem debugging using memory dumps. Since API calls are done in user space, we are interested only in the process and possibly physical memory dumps unless we need to follow syscalls into kernel space. On this slide, I put some general analysis patterns related to function parameters. **Execution Residue** allows us to see past execution traces of Linux API, such as their return addresses in the stack regions. We mention specific memory analysis patterns associated with particular Linux API categories when we look at the latter.

Thread and Adjoint Thread

Before we look at another type of execution history artifacts, trace and log analysis patterns, I illustrate, using the *top* command, the notion of **Adjoint Thread** that figures a lot in trace and log analysis pattern descriptions. In **Thread** (considered abstractly as some sequential activity, here it is USER), we have the same value in its attribute column, and other column values vary. In **Adjoint Thread**, we have the same value for another column, for example, PPID, but the values of other columns, including USER, vary.

Trace and Log Analysis Patterns

◎ Function calls:

- Thread of Activity
- Fiber of Activity
- Adjoint Thread of Activity
- Strand of Activity
- Discontinuity
- Fiber Bundle
- Weave of Activity

© 2023 Software Diagnostics Services

On this slide, I put references to trace and log analysis patterns related to threads, adjoint threads, fiber bundles, and their combinations. For example, discontinuity in messages may show the possible hang when an API function is called.

Thread of Activity

https://www.dumpanalysis.org/blog/index.php/2009/08/03/trace-analysis-patterns-part-7/

Fiber of Activity

https://www.dumpanalysis.org/blog/index.php/2016/06/29/trace-analysis-patterns-part-126/

Adjoint Thread of Activity

https://www.dumpanalysis.org/blog/index.php/2010/03/04/trace-analysis-patterns-part-17/

Strand of Activity

https://www.dumpanalysis.org/blog/index.php/2020/09/12/trace-analysis-patterns-part-198/

Discontinuity

https://www.dumpanalysis.org/blog/index.php/2009/08/04/trace-analysis-patterns-part-8/

Fiber Bundle

https://www.dumpanalysis.org/blog/index.php/2012/09/26/trace-analysis-patterns-part-52/

Weave of Activity

https://www.dumpanalysis.org/blog/index.php/2020/09/19/trace-analysis-patterns-part-201/

API and Errors

- errno.h (values)

- errno

- rtld_errno

- GDB needs an executable linked with –pthread

- FS (x64) and TPIDR_EL0 (ARM64) TLS ABI

Linux API calls may return errors. These are well-documented for individual API functions. However, we may be interested in the overall collection of error values and their meanings since we may find errors in raw stack data (for example, **Execution Residue** memory analysis pattern) and also across different threads. Usually, functions related to syscalls set errno, which has thread-local storage, and to access it from GDB, you need the project linked with the *pthread* library. Minimal loader *libc* implementation uses single-thread rtld_errno.

Values
https://learn.microsoft.com/en-us/cpp/c-runtime-library/errno-constants?view=msvc-170

TLS ABI
https://fuchsia.dev/fuchsia-src/development/kernel/threads/tls

Exercise L7

- **Goal:** Explore error handling implementation in Linux API

- **ADDR Patterns:** Call Epilogue

- **Memory Analysis Patterns:** Last Error Collection

- \LAPI-Dumps\Exercise-L7-GDB.pdf

- \LAPI-Dumps\Exercise-L7-WinDbg.pdf

Exercise L7 (GDB)

Goal: Explore error handling implementation in Linux API.

ADDR Patterns: Call Epilogue.

Memory Analysis Patterns: Last Error Collection.

1.   Load a core dump *core.25399* and *error* executable from the x64 directory:

```
~/LAPI/x64$ gdb -c core.25399 -se error
GNU gdb (Debian 8.2.1-2+b3) 8.2.1
Copyright (C) 2018 Free Software Foundation, Inc.
License GPLv3+: GNU GPL version 3 or later <http://gnu.org/licenses/gpl.html>
This is free software: you are free to change and redistribute it.
There is NO WARRANTY, to the extent permitted by law.
Type "show copying" and "show warranty" for details.
This GDB was configured as "x86_64-linux-gnu".
Type "show configuration" for configuration details.
For bug reporting instructions, please see:
<http://www.gnu.org/software/gdb/bugs/>.
Find the GDB manual and other documentation resources online at:
    <http://www.gnu.org/software/gdb/documentation/>.

For help, type "help".
Type "apropos word" to search for commands related to "word"...
Reading symbols from error...(no debugging symbols found)...done.
[New LWP 25399]
[New LWP 25401]
[Thread debugging using libthread_db enabled]
Using host libthread_db library "/lib/x86_64-linux-gnu/libthread_db.so.1".
Core was generated by `./error'.
#0  0x00007fa7baba75c0 in __GI___nanosleep
(requested_time=requested_time@entry=0x7fffdb5d98e0, remaining=remaining@entry=0x7fffdb5d98e0)
    at ../sysdeps/unix/sysv/linux/nanosleep.c:28
28      ../sysdeps/unix/sysv/linux/nanosleep.c: No such file or directory.
[Current thread is 1 (Thread 0x7fa7baade740 (LWP 25399))]
```

2.   Set logging to a file in case of lengthy output from some commands:

```
(gdb) set logging on L7.log
Copying output to L7.log.
```

3.   Check the *errno* value for the current thread:

```
(gdb) info variables errno
All variables matching regular expression "errno":

File errno.c:
31:     int errno;
27:     int rtld_errno;

File herrno.c:
27:     int __h_errno;
```

163

```
(gdb) p errno
$1 = 2
```

4.   Check the *errno* variable from the other thread:

```
(gdb) info threads
  Id    Target Id                         Frame
* 1     Thread 0x7fa7baade740 (LWP 25399) 0x00007fa7baba75c0 in __GI___nanosleep
(requested_time=requested_time@entry=0x7fffdb5d98e0,
    remaining=remaining@entry=0x7fffdb5d98e0) at ../sysdeps/unix/sysv/linux/nanosleep.c:28
  2     Thread 0x7fa7baadd700 (LWP 25401) 0x00007fa7baba75c0 in __GI___nanosleep
(requested_time=requested_time@entry=0x7fa7baadcea0,
    remaining=remaining@entry=0x7fa7baadcea0) at ../sysdeps/unix/sysv/linux/nanosleep.c:28

(gdb) thread 2
[Switching to thread 2 (Thread 0x7fa7baadd700 (LWP 25401))]
#0  0x00007fa7baba75c0 in __GI___nanosleep
(requested_time=requested_time@entry=0x7fa7baadcea0, remaining=remaining@entry=0x7fa7baadcea0)
    at ../sysdeps/unix/sysv/linux/nanosleep.c:28
28          in ../sysdeps/unix/sysv/linux/nanosleep.c

(gdb) p errno
$2 = 0
```

5.   Check how the variable *errno* is set in the *execve libc* library call:

```
(gdb) disassemble execve
Dump of assembler code for function execve:
   0x00007fa7baba78a0 <+0>:      mov     $0x3b,%eax
   0x00007fa7baba78a5 <+5>:      syscall
   0x00007fa7baba78a7 <+7>:      cmp     $0xfffffffffffff001,%rax
   0x00007fa7baba78ad <+13>:     jae     0x7fa7baba78b0 <execve+16>
   0x00007fa7baba78af <+15>:     retq
   0x00007fa7baba78b0 <+16>:     mov     0xf35b9(%rip),%rcx        # 0x7fa7bac9ae70
   0x00007fa7baba78b7 <+23>:     neg     %eax
   0x00007fa7baba78b9 <+25>:     mov     %eax,%fs:(%rcx)
   0x00007fa7baba78bc <+28>:     or      $0xffffffffffffffff,%rax
   0x00007fa7baba78c0 <+32>:     retq
End of assembler dump.
```

6.   We can also check all per-thread *errno* values at once:

```
(gdb) thread apply all p errno

Thread 2 (Thread 0x7fa7baadd700 (LWP 25401)):
$1 = 0

Thread 1 (Thread 0x7fa7baade740 (LWP 25399)):
$2 = 2
```

Exercise L7 (WinDbg)

Goal: Explore error handling implementation in Linux API.

ADDR Patterns: Call Epilogue.

1.   Launch WinDbg and load a core dump *core.105090* from the A64 directory:

```
Microsoft (R) Windows Debugger Version 10.0.25324.1001 AMD64
Copyright (c) Microsoft Corporation. All rights reserved.


Loading Dump File [C:\LAPI\A64\core.105090]
64-bit machine not using 64-bit API

************* Path validation summary **************
Response                        Time (ms)       Location
Deferred                                        srv*
Symbol search path is: srv*
Executable search path is:
Generic Unix Version 0 UP Free ARM 64-bit (AArch64)
System Uptime: not available
Process Uptime: not available
....
*** WARNING: Unable to verify timestamp for libc.so.6
libc_so+0xb1924:
0000ffff`b8ae1924 d4000001 svc            #0
```

2.   Set the symbol path and logging to a file in case of lengthy output from some commands:

```
0:000> .sympath+ C:\LAPI\A64
Symbol search path is: srv*;C:\LAPI\A64
Expanded Symbol search path is: cache*;SRV*https://msdl.microsoft.com/download/symbols;c:\
lapi\a64


************* Path validation summary **************
Response                        Time (ms)       Location
Deferred                                        srv*
OK                                              C:\LAPI\A64
*** WARNING: Unable to verify timestamp for libc.so.6

0:000> .reload
...*** WARNING: Unable to verify timestamp for libc.so.6
.

************* Symbol Loading Error Summary **************
Module name         Error
libc.so             The system cannot find the file specified

You can troubleshoot most symbol related issues by turning on symbol loading diagnostics (!sym
noisy) and repeating the command that caused symbols to be loaded.
You should also verify that your symbol search path (.sympath) is correct.

0:000> .logopen C:\LAPI\A64\L7-WinDbg.log
Opened log file 'C:\LAPI\A64\L7-WinDbg.log'
```

3.   List threads and their stack traces:

```
0:000> ~*k
```

```
Unable to get thread data for thread 0
.  0  Id: 19a82.19a82 Suspend: 0 Teb: 00000000`00000000 Unfrozen
 # Child-SP          RetAddr            Call Site
00 0000ffff`e57823a0 0000ffff`b8ae6aec  libc_so!clock_nanosleep+0x104
01 0000ffff`e5782420 0000ffff`b8ae69b8  libc_so!_nanosleep+0x1c
02 0000ffff`e5782430 0000aaaa`d9be0950  libc_so!sleep+0x48
03 0000ffff`e5782480 0000ffff`b8a573fc  error!main+0x5c
04 0000ffff`e57824b0 0000ffff`b8a574cc  libc_so!_libc_init_first+0x7c
05 0000ffff`e57825c0 0000aaaa`d9be07f0  libc_so!_libc_start_main+0x98
06 0000ffff`e5782620 ffffffff`ffffffff  error!start+0x30
07 0000ffff`e5782620 00000000`00000000  0xffffffff`ffffffff

Unable to get thread data for thread 1
   1  Id: 19a82.19a83 Suspend: 0 Teb: 00000000`00000000 Unfrozen
 # Child-SP          RetAddr            Call Site
00 0000ffff`b8a2e740 0000ffff`b8ae6aec  libc_so!clock_nanosleep+0x104
01 0000ffff`b8a2e7c0 0000ffff`b8ae69b8  libc_so!_nanosleep+0x1c
02 0000ffff`b8a2e7d0 0000aaaa`d9be08e8  libc_so!sleep+0x48
03 0000ffff`b8a2e820 0000ffff`b8aad5c8  error!thread_func+0x14
04 0000ffff`b8a2e840 0000ffff`b8b15d1c  libc_so!pthread_condattr_setpshared+0x4f8
05 0000ffff`b8a2e960 ffffffff`ffffffff  libc_so!clone+0x5c
06 0000ffff`b8a2e960 00000000`00000000  0xffffffff`ffffffff
```

4.   Check how the variable *errno* is set in the *execve libc* library call:

```
0:000> uf execve
libc_so!_libc_start_main+0x15c:
0000ffff`b8a57590 90000ba2 adrp      x2,libc_so!sys_siglist+0x1a8 (0000ffff`b8bcb000)
0000ffff`b8a57594 f946e442 ldr       x2,[x2,#0xDC8]
0000ffff`b8a57598 d53bd043 mrs       x3,TPIDR_EL0
0000ffff`b8a5759c aa0003e1 mov       x1,x0
0000ffff`b8a575a0 92800000 mov       x0,#-1
0000ffff`b8a575a4 4b0103e1 neg       w1,w1
0000ffff`b8a575a8 b8226861 str       w1,[x3,x2]
0000ffff`b8a575ac d65f03c0 ret

libc_so!execve:
0000ffff`b8ae74c0 d503201f nop
0000ffff`b8ae74c4 d2801ba8 mov       x8,#0xDD
0000ffff`b8ae74c8 d4000001 svc       #0
0000ffff`b8ae74cc b13ffc1f cmn       x0,#0xFFF
0000ffff`b8ae74d0 54000042 bhs       libc_so!execve+0x18 (0000ffff`b8ae74d8)  Branch

libc_so!execve+0x14:
0000ffff`b8ae74d4 d65f03c0 ret  Branch

libc_so!execve+0x18:
0000ffff`b8ae74d8 17fdc02e b         libc_so!_libc_start_main+0x15c (0000ffff`b8a57590)
Branch
```

API and Functional Programming

- <u>Referential transparency</u>

- strlen

      strlen(s), strlen(s) → n, n

- read

      read(fd, buf, len),
      read(fd, buf, len) → n1, n2

- <u>Side effects</u>

With the rise of functional programming during the last decade and due to my own experience with FP, I put this new that shows the relation of API to referential transparency and side effects. In summary, referential transparency allows the substitution of expressions with their value. This requires that the result of the expression is the same for the same input values and that there are no side effects, modifications of memory, or I/O outside of the expression. For example, the length of the string is the same for the same string, and no memory is modified outside. In contrast, the *read* function not only may have a different output for the same parameter values but also modifies memory outside with unknown values every time it is called.

Referential transparency
https://en.wikipedia.org/wiki/Referential_transparency

Side effects
https://en.wikipedia.org/wiki/Side_effect_(computer_science)

API and Security

- ◉ Maliciousness: What, When, Where

  - SecQuant: Quantifying Container System Call Exposure
  - System call risk level classification / syscalls / masks

- ◉ Vulnerability: How

  - SAST
  - Static code analysis tools

It is difficult to put all security issues with API on just one slide. So, I attempted to make a view of the surface from a distant star. There are two main aspects: maliciousness of API usage, for example, syscalls, and how the API is used to make it vulnerable, for example, runtime API libraries.

SecQuant: Quantifying Container System Call Exposure
http://www.cs.toronto.edu/~sahil/suneja-esorics22.pdf

System call risk level classification
https://wims.univ-cotedazur.fr/sysmask/doc/technical_txt/risklevels.txt

syscalls
https://wims.univ-cotedazur.fr/sysmask/doc/syscalls-bynumber.html

masks
https://wims.univ-cotedazur.fr/sysmask/doc/masks.txt

SAST
https://en.wikipedia.org/wiki/Static_application_security_testing

Static code analysis tools
https://en.wikipedia.org/wiki/List_of_tools_for_static_code_analysis

API and Versioning

- Windows: Ex-suffix, longer descriptive function names

  CopyFile/CopyFileEx
  CreateThread/CreateRemoteThread

- Linux: numbering, shorter prefixes/suffixes

  openat/openat2
  accept/accept4
  read/pread/readv/preadv/preadv2

A few words about API versioning in case of adding extensions. In Windows, this is often done via the Ex-suffix or by creating more longer descriptive names. In Linux, it is often done by adding a number or short prefixes and suffixes.

Linux API Formalization

Linux API Formalization

Ideas from Conceptual Mathematics

Now we have a break. We look at possible API formalization using ideas from conceptual mathematics, such as category theory. The exposition is informal. If you already know category theory, you find another application or even provide objections to the approach if you see it wrong. However, if you don't know category theory, you learn its basic notions and terminology and perhaps apply it to your own areas of interest.

API Compositionality

- ⊙ <u>Principle of compositionality</u>

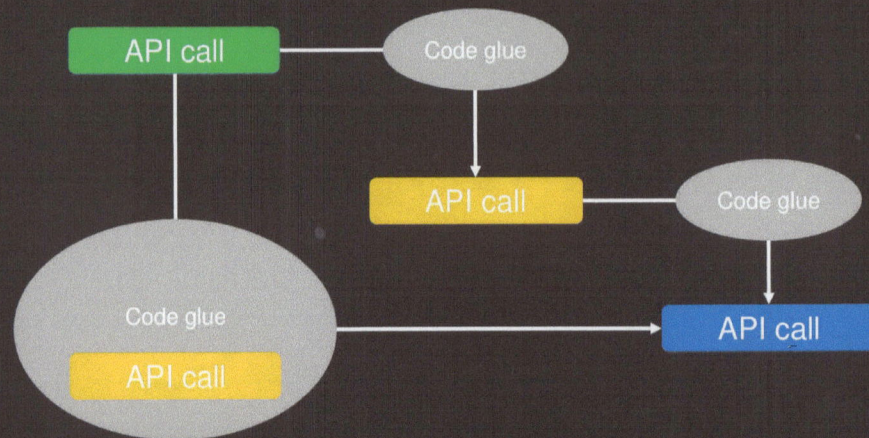

© 2023 Software Diagnostics Services

The main principle that allows us to use category theory that I introduce next is the so-called principle of compositionality. We can compose various API calls together via code glue. There's also possible to connect two API calls by some code, including other API calls if necessary. Here we represent API calls as some objects and code glue as directed arrows.

Principle of compositionality
https://en.wikipedia.org/wiki/Principle_of_compositionality

Category Theory Language

- ⊙ Category

  - Objects
  - Arrows between objects (must be transitive, if A → B and B → C then A → C)

- ⊙ Functor

  - Arrow between categories (can be the same category)
  - Maps objects to objects and arrows to arrows

- ⊙ Natural Transformation

  - Arrows between functors in a category of functors

- ⊙ Adjunction

  - Relationship between functors, change of perspective, back translation

Category theory allows us to relate different areas analogically via their common structure and behavior. It was introduced in mathematics in the middle of the 20th century. And it has its own language. The three main concepts are **categories** themself, **functors**, and **natural transformations**. To this, I add the so-called **adjunction**. The definitions I give here are all informal but suitable for our application to API. A category consists of objects and arrows between objects. However, arrows, if they exist between objects, must be composable. There are other restrictions and axioms that I omit for our informal purposes. Categories themselves can be considered objects. Arrows between them are called functors. Functors map objects to objects and arrows to arrows between source and target categories. Functors can be considered as objects themselves in a category of functors. Arrows between them are called natural transformations. The 4th concept, adjunction, is a pair of functors, called left and right adjoints, that are in a special relationship to each other that allows changing of perspective when traveling back and forth between categories. We now make all these 4 concepts visual on the next slide.

A View of Category Theory

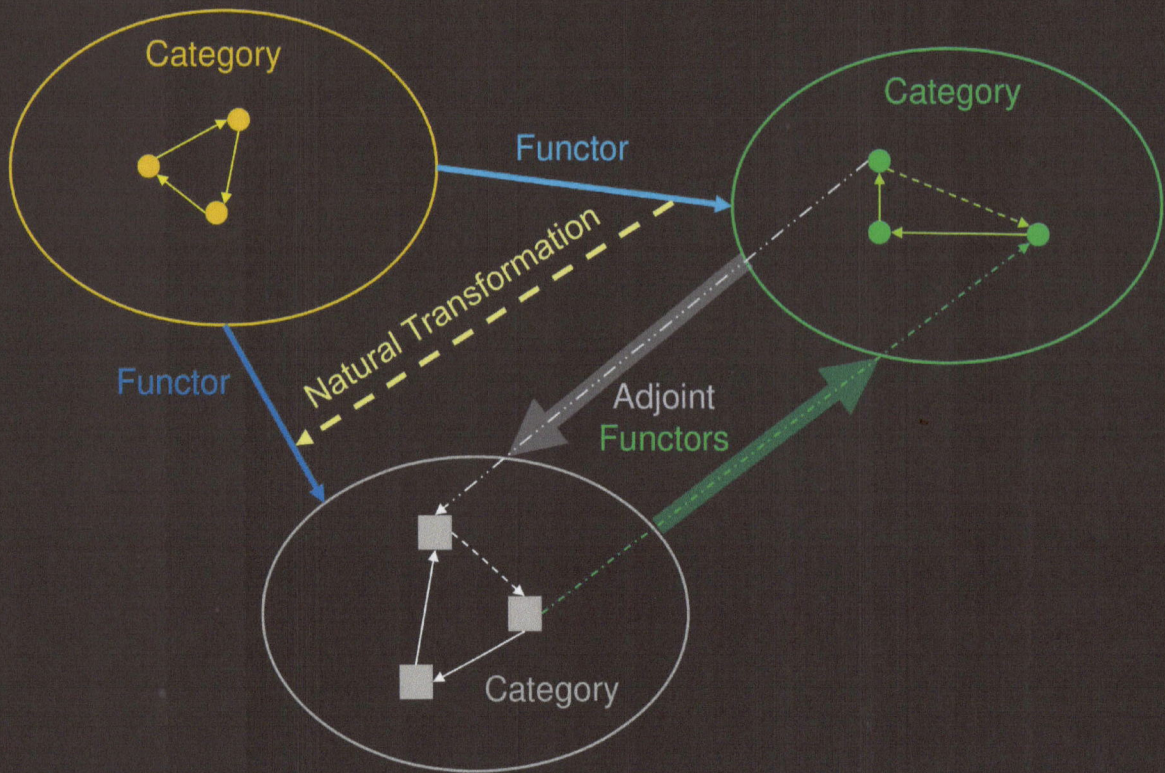

This slide is a visual overview of the four main concepts: category, functor between categories, natural transformation between functors, and adjoint functors. Please notice a kind of tunnel functor arrows that highlight back translation between objects of categories.

Category Theory Square

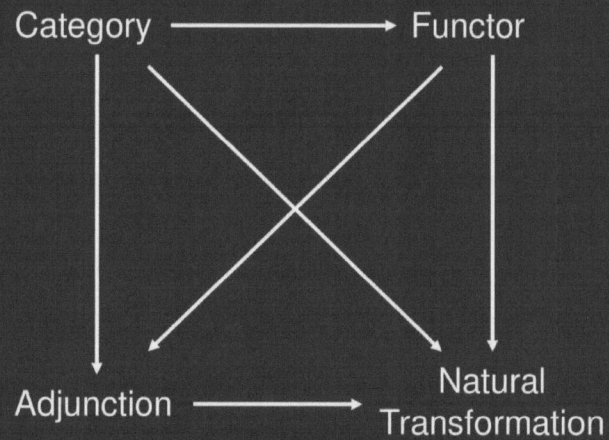

We can arrange all four concepts via the **category theory square** that shows their relationship that is easy to remember.

API Category

- API as objects, glue code as arrows

- API as arrows, glue code as objects fails at composition

- Initial and terminal API objects in subcategories

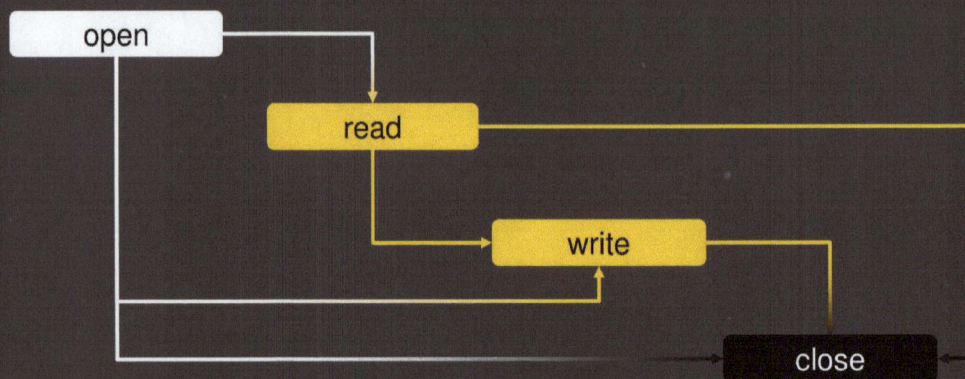

Let's translate all those informal abstract concepts into the API world. Again, in the API category, API functions are objects, and glue code are arrows. The other possible dual arrangement, API as arrows, is invalid due to impossible compositions: we cannot get API by composing two API calls. The result of the composition is an object, glue code. Some API functions can be initial or terminal in small subcategories, for example, *open* and *close*.

API Functor

- Translates between API layers (different API)

- Stack trace as functor

- Translates between different API sequences

- Endofunctor – between the same API

- Translates between different code implementations

Different API layers could be different API categories or subcategories of the same category. All such translations are functors. It also suggests interpreting stack traces as functors. Finally, translation between different API sequences is also a functor that maps API calls to API calls and glue code to glue code.

API Diagram

- Indexed set $\rightarrow$ diagram

- Functor from a shape (pattern)

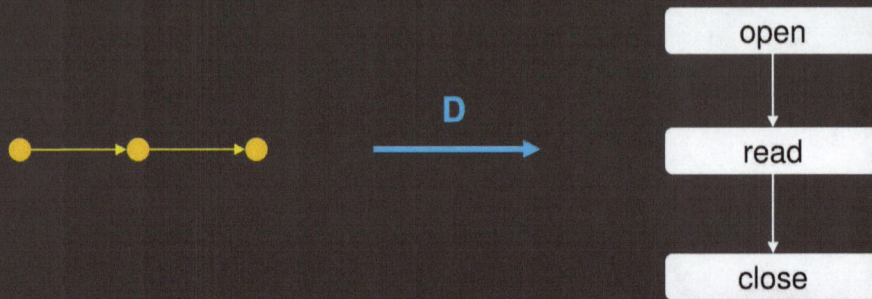

You may have noticed that category theory illustrations are similar to graphs and diagrams. In category theory, a diagram in some category **C** is a functor D from some shape category where we don't care about individual objects and arrows to that category **C**. They are categorical analogs to indexed sets where a shape is a set of indexes.

API Natural Transformation

- Maps between different vertical API sequences (stack traces)

- Maps between different code translations

- Diagnostics and debugging as natural transformation

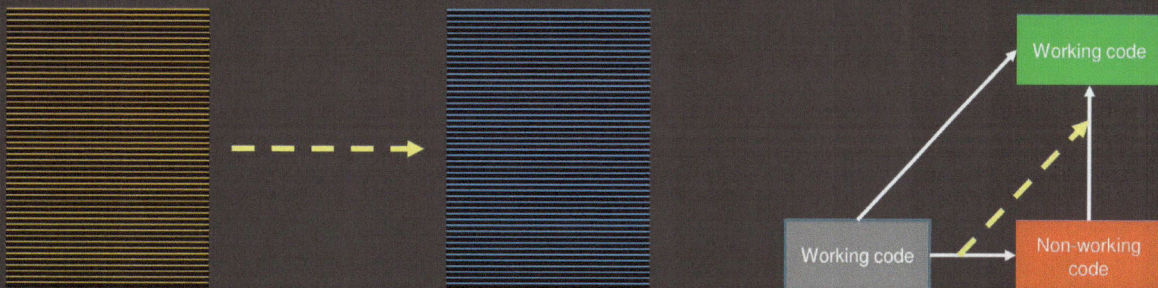

If stack traces are functors, then maps between them are natural transformations. The same goes for maps between different code translations. Hence, it naturally suggests a categorical interpretation of diagnostics and debugging as natural transformation.

Cross-platform API

⊙ Windows API / Linux API

⊙ Similar diagrams

⊙ Cross-platform development as a natural transformation

Diagrams are functors. Therefore, maps between diagrams are natural transformations. We can represent various API sequences in different operating systems as similar diagrams, and this enables us to view cross-platform development as a natural transformation.

API Adjunction

- Navigation between different API sequences

- Call and return stack trace sequences, callbacks (when stack traces correspond to vertical API sequences)

- Back translation between traces/logs (when traces correspond to API horizontal sequences)

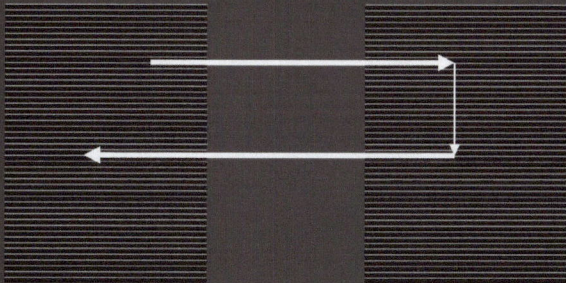

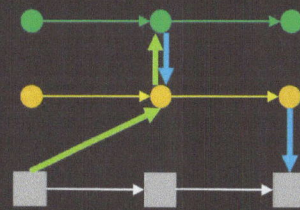

API adjunction formalizes back-and-forth navigation between different API sequences, be it traces and logs (horizontal API sequences) or call and return stack trace sequences (vertical API sequences). In some way, it corresponds to changes in perspective, especially when navigating around logs.

Informal n-API

- Arrows between arrows

- 1-API – normal API usage

- 2-API – diagnostics, debugging

- 3-API – higher diagnostics, debugging (debugging the debugging)

- ∞-API – for homework ☺

On this slide, we suggest the further informal application of n-category theory, where we consider arrows between arrows. 1-category here maps to normal API usage. 2-category corresponds to diagnostics and debugging. 3-category corresponds to higher level diagnostics and debugging, diagnosing diagnostics and debugging, debugging diagnostics and debugging.

API and Trace Categories

- 1-category API ([semigroup](#))

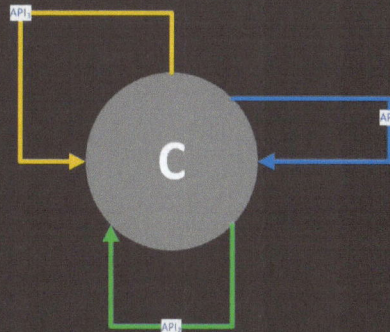

- [2-category of traces and logs](#)

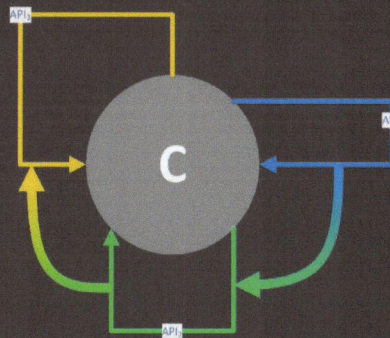

There can be various categories formed from API and code. We have already looked at one where the API calls are objects connected via code arrows. There can be a different view and a different category of one object, Code, and API calls as arrows. Arrows between arrows may correspond to traces and logs, and these form the 2-category. The original motivation for this view comes from this link:

2-category of traces and logs
https://www.dumpanalysis.org/traces-logs-as-2-categories

semigroup
https://en.wikipedia.org/wiki/Semigroup

API I/O

- Categories – one input, one output

- Operads – many inputs, one output

- Properads – many inputs, many outputs

```
int socketpair(
        int domain,
        int type,
        int protocol,
        int sv[2]
);
```

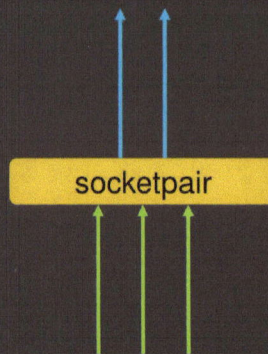

socketpair

Finally, we consider API functions in terms of their input/output. Since objects of categories have one input and one output, we abstract from API input and consider it as some glue code. In functional programming languages, there may be many inputs, and one produced output, and these mathematical objects (operations) are called operads. In Linux API, outputs may be modified, producing new values, and therefore, we have several inputs and outputs, and these mathematical operations are called properads. At the end of the training, I put a slide for further reading.

Linux API and Languages

Linux API and Languages

Now we look at how Linux API is used by various languages other than C and C++. The choice of languages was dictated by my own interests. Fortunately, all of them are mainstream. I've chosen the simple example of calling the *execve* syscall wrapper function.

189

API and C#

- Installation

- Platform Invoke

Our first language is C#. The way to call Linux API is to use the so-called **P/Invoke** mechanism. Since .NET and C# are not the main Linux languages, I created an exercise with the appropriate installation steps.

Installation
https://learn.microsoft.com/en-us/dotnet/core/install/linux

Platform Invoke
https://learn.microsoft.com/en-us/dotnet/standard/native-interop/pinvoke

Exercise L8

- **Goal:** Install .NET environment and write a simple program that uses Linux API

- \LAPI-Dumps\Exercise-L8.pdf

© 2023 Software Diagnostics Services

Exercise L8

Goal: Install .NET environment and write a simple program that uses Linux API.

1.   Install .NET SDK on your distribution (https://learn.microsoft.com/en-us/dotnet/core/install/linux). We use Debian 10 on WSL for demonstration (https://learn.microsoft.com/en-us/dotnet/core/install/linux-debian).

```
~/LAPI$ wget https://packages.microsoft.com/config/debian/10/packages-microsoft-prod.deb -O
packages-microsoft-prod.deb

~/LAPI$ sudo dpkg -i packages-microsoft-prod.deb

~/LAPI$ rm packages-microsoft-prod.deb

~/LAPI$ sudo apt-get update

~/LAPI$ sudo apt-get install -y dotnet-sdk-7.0
```

2.   Verify that .NET SDK and runtime are installed.

```
~/LAPI$ dotnet --list-sdks
7.0.302 [/usr/share/dotnet/sdk]

~/LAPI$ dotnet --list-runtimes
Microsoft.AspNetCore.App 7.0.5 [/usr/share/dotnet/shared/Microsoft.AspNetCore.App]
Microsoft.NETCore.App 7.0.5 [/usr/share/dotnet/shared/Microsoft.NETCore.App]
```

3.   Create SBT native project:

```
~/LAPI$ mkdir csharp-lapi

~/LAPI$ cd csharp-lapi

~/LAPI/csharp-lapi$ dotnet new console

Welcome to .NET 7.0!
---------------------
SDK Version: 7.0.302

Telemetry
---------
The .NET tools collect usage data in order to help us improve your experience. It is collected
by Microsoft and shared with the community. You can opt-out of telemetry by setting the
DOTNET_CLI_TELEMETRY_OPTOUT environment variable to '1' or 'true' using your favorite shell.

Read more about .NET CLI Tools telemetry: https://aka.ms/dotnet-cli-telemetry

----------------
Installed an ASP.NET Core HTTPS development certificate.
To trust the certificate run 'dotnet dev-certs https --trust' (Windows and macOS only).
Learn about HTTPS: https://aka.ms/dotnet-https
----------------
Write your first app: https://aka.ms/dotnet-hello-world
Find out what's new: https://aka.ms/dotnet-whats-new
Explore documentation: https://aka.ms/dotnet-docs
```

```
Report issues and find source on GitHub: https://github.com/dotnet/core
Use 'dotnet --help' to see available commands or visit: https://aka.ms/dotnet-cli
----------------------------------------------------------------------------
The template "Console App" was created successfully.

Processing post-creation actions...
Restoring /home/coredump/LAPI/csharp-lapi/csharp-lapi.csproj:
  Determining projects to restore...
  Restored /home/coredump/LAPI/csharp-lapi/csharp-lapi.csproj (in 142 ms).
Restore succeeded.
```

4.   Open *Program.cs* and replace code with:

```
using System.Runtime.InteropServices;

class LapiTest
{
  [DllImport("libc")]
  public static extern int execve(string __path, string[] __argv, string[] __envp);

  public static void Main()
  {
    execve("/bin/ps", new string[0], new string[0]);
  }
}
```

5.   Run the project:

```
~/LAPI/csharp-lapi/lapi$ dotnet run
    PID TTY          TIME CMD
  37187 pts/2    00:00:00 bash
  44981 pts/2    00:00:06 dotnet
  45267 pts/2    00:00:02 dotnet
  45300 pts/2    00:00:00 ps
```

API and Scala Native

- Documentation

- Scala Native

- Native code interoperability

My second language choice was Scala, the language I started learning more than 3 years ago. It was originally created for JVM, but there's also a native implementation using LLVM. Since the choice of language is rather unusual, I created a special exercise for it that also details the required Scala Native installation steps. Over the last 3 years, Scala Native has really improved.

Documentation
https://buildmedia.readthedocs.org/media/pdf/scala-native/latest/scala-native.pdf

Scala Native
https://scala-native.org/en/stable/index.html

Native code interoperability
https://scala-native.org/en/stable/user/interop.html

Exercise L9

- **Goal:** Install Scala Native environment and write a simple program that uses Linux API

- \LAPI-Dumps\Exercise-L9.pdf

Exercise L9

Goal: Install Scala Native environment and write a simple program that uses Linux API.

1.   Install JDK.

```
~/LAPI$ sudo apt update
```

```
~/LAPI$ sudo apt install openjdk-11-jdk
```

```
~/LAPI$ java --version
openjdk 11.0.18 2023-01-17
OpenJDK Runtime Environment (build 11.0.18+10-post-Debian-1deb10u1)
OpenJDK 64-Bit Server VM (build 11.0.18+10-post-Debian-1deb10u1, mixed mode, sharing)
```

2.   Install Scala with cs setup as recommended at https://www.scala-lang.org/download/.

```
~/LAPI$ curl -fL https://github.com/coursier/coursier/releases/latest/download/cs-x86_64-pc-
linux.gz | gzip -d > cs && chmod +x cs && ./cs setup
  % Total    % Received % Xferd  Average Speed   Time    Time     Time  Current
                                 Dload  Upload   Total   Spent    Left  Speed
  0     0    0     0    0     0      0      0 --:--:-- --:--:-- --:--:--     0
  0     0    0     0    0     0      0      0 --:--:-- --:--:-- --:--:--     0
100 19.7M  100 19.7M    0     0  10.6M      0  0:00:01  0:00:01 --:--:-- 27.6M
Checking if a JVM is installed
Found a JVM installed under /usr/lib/jvm/java-11-openjdk-amd64.

Checking if ~/.local/share/coursier/bin is in PATH
  Should we add ~/.local/share/coursier/bin to your PATH via ~/.profile? [Y/n]

Checking if the standard Scala applications are installed
  Installed ammonite
  Installed cs
  Installed coursier
  Installed scala
  Installed scalac
  Installed scala-cli
  Installed sbt
  Installed sbtn
  Installed scalafm
```

```
~/LAPI$ scala -version
Scala code runner version 3.3.0 -- Copyright 2002-2023, LAMP/EPFL
```

Note: You may need to log out and log in to make installation changes into effect.

3.   Install *clang*:

```
~/LAPI$ clang --version
clang version 7.0.1-8+deb10u2 (tags/RELEASE_701/final)
Target: x86_64-pc-linux-gnu
Thread model: posix
InstalledDir: /usr/bin
```

4. Create SBT native project:

```
~/LAPI$ mkdir scala-lapi

~/LAPI$ cd scala-lapi

~/LAPI/scala-lapi$ sbt new scala-native/scala-native.g8
...
A minimal project that uses Scala Native.

name [Scala Native Seed Project]: lapi

Template applied in /home/coredump/LAPI/scala-lapi/./lapi

~/LAPI/scala-lapi$ cd lapi
```

5. Open \src\main\scala\Main.scala and replace code with:

```scala
import scala.scalanative.unsafe._

@extern
object libc {
  def execve(__path: CString, __argv: Ptr[CString], __envp: Ptr[CString]): Int = extern
}

object Main {
  def main(args: Array[String]): Unit =
    libc.execve(c"/bin/ps", 0.toPtr, 0.toPtr)
}
```

6. Run the project from sbt:

```
~/LAPI/scala-lapi/lapi$ sbt run
[info] welcome to sbt 1.8.3 (Debian Java 11.0.18)
[info] loading settings for project lapi-build from plugins.sbt ...
[info] loading project definition from /home/coredump/LAPI/scala-lapi/lapi/project
[info] loading settings for project lapi from build.sbt ...
[info] set current project to lapi (in build file:/home/coredump/LAPI/scala-lapi/lapi/)
[info] compiling 1 Scala source to
/home/coredump/LAPI/scala-lapi/lapi/target/scala-3.2.2/classes ...
[info] Linking (1586 ms)
[info] Discovered 671 classes and 3742 methods
[info] Optimizing (debug mode) (3037 ms)
[info] Generating intermediate code (1352 ms)
[info] Produced 8 files
[info] Compiling to native code (3526 ms)
[info] Total (9745 ms)
    PID TTY          TIME CMD
  37187 pts/2    00:00:00 bash
  41051 pts/2    00:01:26 java
  41916 pts/2    00:00:00 ps
[success] Total time: 16 s, completed May 31, 2023, 10:39:08 PM
```

7. Run the native executable:

```
~/LAPI/scala-lapi/lapi$ ./target/scala-3.2.2/lapi-out
    PID TTY          TIME CMD
  37187 pts/2    00:00:00 bash
  43076 pts/2    00:00:00 ps
```

197

API and Golang

- <u>unix</u>

- Example:

```
package main

import (
        "golang.org/x/sys/unix"
)

func main() {
        unix.Exec("/bin/ps", nil, nil)
}
```

Another popular language choice is Golang, which I recently used in conjunction with C++ for some projects. There's a unix package, and its simple usage is illustrated on this slide. I haven't created an explicit exercise for this mainstream Linux language for cloud environments.

unix package
https://pkg.go.dev/golang.org/x/sys/unix

API and Rust

- Unsafe: libc

- Safe: nix, rustix

- Example:

```
use libc::{c_char, execve};

fn main() {
    unsafe {
        execve("/bin/ps".as_ptr() as *const c_char, std::ptr::null(), std::ptr::null());
    }
}
```

Another of my favorite languages is Rust. The example here uses the unsafe crate *libc*. Other crates are safe wrappers. I haven't created an explicit exercise for this language, so please follow the official language installation documentation.

libc
https://lib.rs/crates/libc

nix
https://lib.rs/crates/nix

rustix
https://lib.rs/crates/rustix

API and Python

- **os**

- **ctypes**

- Example:

```
import ctypes

ctypes.CDLL("libc.so.6").execve("/bin/ps".encode("utf-8"), 0, 0)
```

And finally, Python, where the *ctypes* library can be used if the higher level os package is not sufficient. The code is the simplest among all examples. The provided library link documentation has additional examples. I haven't created an explicit exercise for this mainstream Linux language because, usually, Python interpreter is already included in many Linux distributions.

os library
https://docs.python.org/3/library/os.html

ctypes library
https://docs.python.org/3/library/ctypes.html

Linux API Classes

Linux API Classes

The final section of this training is a tour through selected Linux API classes. I don't use the term category to avoid confusion with category theory. I chose API classes based on my experience with monitoring and also reading various Linux API books. Most of the classes have two sections: Library and Syscalls. Library API may use the corresponding syscalls in their implementation. Most likely, the second edition of this course will have more classes added. The goal of this section is to give a bird's eye overview of API possibilities to stimulate further study and perhaps fill some gaps if you have.

Tracing

- <u>strace</u> (syscalls, signals)

- <u>ltrace</u> (libraries, syscalls, signals)

- Examples

  - strace bash
  - ltrace bash
  - ltrace -S bash

There are two tools that help in seeing the dynamic usage of Linux API in processes, both library and system calls. The *ltrace* tool can log both types.

strace
https://man7.org/linux/man-pages/man1/strace.1.html

ltrace
https://man7.org/linux/man-pages/man1/ltrace.1.html

Classification

- Classification and Grouping of Linux System Calls

- A study of modern Linux API usage and compatibility (slides)

- A study of modern Linux API usage and compatibility (paper)

There were a few attempts to classify Linux API, primarily syscalls.

Classification and Grouping of Linux System Calls
http://seclab.cs.sunysb.edu/sekar/papers/syscallclassif.htm

A study of modern Linux API usage and compatibility (slides)
https://oscarlab.github.io/api-compat-study/files/syspop-eurosys16-slides.pdf

A study of modern Linux API usage and compatibility (paper)
https://www.chiachetsai.com/files/eurosys16.pdf

System Configuration API

- ⊙ Library

  - Invariant and increasable `sysconf`
  - Pathnames `(f)pathconf`

- ⊙ Syscalls

  - Identification `uname`

System configuration API allows querying various properties and information about the kernel, such as version.

File I/O API

- ◉ [File I/O Essentials](#)

- ◉ Layers

  - Stream-based   `f(open|seek|read|write|close)`   `FILE*`
  - Syscall-based   `creat|open|lseek|read|write|ioctl|close fd`

- ◉ Buffering

  - Block-based   `f(read|write)`
  - Line-based   `(f)puts, (f)printf`
  - Unbuffered   `(p)read(v), (p)write(v)`

- ◉ Multiplexing   `(p)select, poll, epoll_(create|ctl|wait)`

Since everything (or almost everything) is a file in Linux, we start with File I/O. I provided a link to the excellent free chapter on such API with many hands-on examples. Besides the usual split into library and syscall layers, we can architecturally view them as stream and file-descriptor based. Another API classification is possible when we take into account buffering. For line-based buffering examples, I provided only two examples, and there are many more functions. It is also possible to do multiplexing, waiting for data from many file descriptors at once.

File I/O Essentials
https://static.packt-cdn.com/downloads/File_IO_Essentials.pdf

File Control API

- ⊙ Library

  - Mixing `FILE*` and `fd fileno`, `fdopen`
  - Buffering `set(v)buf`, `fflush`
  - Temp `mkstemp`, `tmpfile`

- ⊙ Syscalls

  - Buffering `f(data)sync`
  - Access `posix_fadvise`
  - Control `fcntl`
  - Locking `flock`
  - Truncation `(f)truncate`

File control API allows changing the behavior of file I/O, such as set buffering parameters, and also provides conversion between file streaming structure and file descriptor. I also included functions for making temporary file names and files.

Filesystem API

- ⊙ Library

  - Directories `(n)ftw`, `(fd)(open|read)dir`, `remove`, `get(c)wd`
  - Paths `realpath`, `(dir|base)name`

- ⊙ Syscalls

  - Mounting `(u)mount`
  - Metadata `(f|l)stat(at|vfs)`
  - Attributes `(list|get|set)(f|l)xattr`
  - Permissions `(l|f)ch(own|mod)`, `umask`
  - Time `utime(s)`
  - Links `(un)link`, `rename`
  - Symlinks `(sym|read)link`
  - Directories `(mk|rm|ch)dir`, `chroot`
  - Monitoring `inotify_(init|(add|rm)_watch)`

Files are organized into filesystem hierarchies. There are several library functions and syscalls to mount filesystems, modify metadata, extended attributes, file and directory permissions, get file access and modification times, create hard and soft links, and do file activity monitoring.

Dynamic Memory API

- ◉ Library

  - Control `mall(opt|info)`, `memalign`
  - Allocation `(m|c|re)alloc`
  - Deallocation `free`
  - Debugging `m(un)trace`, `m(check|probe)`
  - Stack `alloca`

- ◉ Syscalls

  - Adjustment `(s)brk`

In this dynamic memory class, I included both heap and stack. Heap API is the most famous API from a software diagnostics perspective. The syscall only manages the heap region as a whole. The actual heap dynamic memory implementation is entirely in the library. Please note various useful debugging library functions. For small dynamic arrays and structures, it is sometimes faster to use local stack memory.

Virtual Memory API

- ⊙ Library

  - Usage `posix_madvise`

- ⊙ Syscalls

  - Mapping `m((un|re)map|sync)`, `remap_file_pages`
  - Protection `mprotect`
  - Locking `m(un)lock(all)`
  - Residence `mincore`
  - Usage `madvise`

Virtual memory API allows mapping files to memory for faster access and allocation of memory regions with different protections.

Shared Libraries API

- Library

  - Loading dlopen
  - Unloading dlclose
  - Errors dlerror
  - Symbols dl(addr|sym)

Shared library API uses virtual memory API to load .so files. It also includes functions to query addresses of symbols, for example, exported functions.

Process API

- ◉ Library

  - Execution `exec(v(p)|l(e|p))`, `fexecve`, `system`
  - Termination `exit`
  - Exit `(at|on_)exit`
  - Capabilities `cap_((get_|set_)proc|free)`

- ◉ Syscalls

  - Creation `fork`, `clone`
  - Execution `execve`
  - Termination `_exit`
  - Waiting `wait(pid|id|3|4)`
  - Resources `acct`, `getrusage`, `(get|set)rlimit`
  - Priority `(get|set)priority`, `sched_get_priority_(min|max)`
  - Scheduling `sched_((set|get)(scheduler|param)|yield|rr_get_interval)`
  - Affinity `sched_(set|get)affinity`
  - Capabilities `prctl`

Process API includes functions to create, execute, and terminate processes, as well as to wait for them to finish their execution. The clone syscall allows for fine-grain control over process creation than just fork. Additional functions exist to set termination handlers, get resource usage and set their resource limits, priority, scheduling parameters, CPU affinity, and capabilities.

IPC API

- ⊙ Library

  - Pipes `p(open|close)`   `FILE*`
  - FIFO `mkfifo`
  - Keys `ftok`
  - POSIX `(mq|sem|shm)_(open|unlink|*)`

- ⊙ Syscalls

  - Pipes `pipe`
  - Handles `dup(2)`, `close`
  - Message queues `msg(get|snd|rcv|ctl)`
  - Semaphores `sem(get|ctl|op)`
  - Shared memory `shm(get|at|ctl|dt)`

Inter-process communication API includes functions for pipes, queues, shared memory, and semaphores for synchronization.

Job API

- ⊙ Library

  - Terminal `ctermid`, `tc(get|set)pgrp`

- ⊙ Syscalls

  - Groups `(get|set)pgid`
  - Sessions `(get|set)sid`

Processes may be organized into groups and groups into sessions. Some processes may have controlling terminals for console I/O.

Signals API

- ⊙ Library

  - Sets `sig(((empty|fill|add|del|and|or|isempty)set)|ismember)`
  - Sending `raise, killpg, abort`
  - Description `strsignal, psignal`
  - Waiting `pause`
  - BSD `sig(vec|block|(get|set)mask|pause)`

- ⊙ Syscalls

  - Disposition `signal, sigaction`
  - Mask `sig(procmask|pending)`
  - Sending `kill, sigqueue`
  - Waiting `sig(suspend|waitinfo), signalfd`
  - Stack `sigaltstack`

Signals are also a way for interprocess communication as well as notification when something illegal happens, like invalid memory access. They are conceptually similar to software interrupts and exceptions with the possibility to set a separate stack for processing.

Thread API

- ◉ Library

  - One-time initialization `pthread_once`
  - Creation `pthread_(create|atfork)`
  - Termination `pthread_exit`
  - Identification `pthread_(self|equal)`
  - Wait `pthread_(join|detach)`
  - Cancellation `pthread_(set|test)cancel(state|type)`
  - Cleanup `pthread_cleanup_(push|pop)`
  - Signals `pthread_(sigmask|kill|sigqueue), sigwait`
  - Attributes `pthread_attr_(init|set*|destroy)`
  - Synchonization `pthread_mutex(_(init|(un|try|timed)lock|destroy)| attr_(init|set*|destroy)) pthread_cond_(init|signal|broadcast|(timed)wait|destroy)`
  - Data `pthread_(key_create|(set|get)specific)`

Threads (or lightweight processes) have their own library API. Internally, it uses process API syscalls. This library also includes mutexes and conditional variables.

Networking API

- ⊙ Library

  - Addresses `(get|free)addrinfo, gai_strerror`
  - Names `getnameinfo`

- ⊙ Syscalls

  - Sockets `socket, bind, listen, accept, connect, close, shutdown`
  - Streaming `write, send, read, recv`
  - Datagrams `recvfrom, sendto`
  - Messages `(send|recv)msg`
  - Files `sendfile`
  - Domain `socketpair`
  - Addresses `get(sock|peer)name`
  - Options `(get|set)sockopt`

Networking API includes UNIX domain and TCP/IP sockets and functions to get address information. When preparing this slide, I myself found a few functions I didn't know about.

Time API

- ⊙ **Library**

  - Conversion `(c|asc|strf|strp|get|local|mk)time`
  - Zones `tzset`
  - Locales `setlocale`
  - Correction `adjtime`
  - Process `clock`

- ⊙ **Syscalls**

  - Calendar `(get|set)timeofday, (s)time`
  - Process `times`

Time API includes various conversion functions and also syscalls to get calendar system time and process times spent in user and kernel modes.

Timers API

- ⦿ Library
  - Low-res `sleep`
  - Clock (`clock|pthread`)`_getcpuclockid`

- ⦿ Syscalls
  - Intervals `(get|set)itimer`, `timer_(create|settime|delete)`
  - Repeating `timer_getoverrun`
  - Once `alarm`
  - File `timerfd_(create|(get|set)time)`
  - Clock `clock_(get|set)(time|res)`
  - Hi-res `(clock_)nanosleep`

Programs need some sleep and also alarm clocks.

Tracing and Logging API

- ⊙ Library

  - Writing `(open|(v)sys|close)log`
  - Filtering `setlogmask`

- ⊙ Syscalls

  - Process `ptrace`
  - Performance and monitoring `perf_event_open`
  - eBPF `bpf`

There are several ways to trace syscalls and monitor performance. The oldest one is perf events API, and the more recent is eBPF API. System logging is implemented at the library level.

Accounts API

- ⊚ Library

  - Records `getpw(nam|uid)`, `getgr(nam|gid)`
  - Scanning `getspnam`, `(set|get|end)spent`
  - Groups `(init|get|set)groups`
  - Encryption `crypt(_r)`

- ⊚ Syscalls

  - ID `(get|set)((r)e)(s)(u|g)id`
  - Filesystem `setfs(u|g)id`

The accounts API class helps get user and group information, get and set their identity, as well as do password encryption.

Terminal API

- ⦿ Library

  - Identification `isatty`, `ttyname`
  - Attributes `tc(get|set)attr`
  - Speed `cf(get|set)(to|ti)speed`
  - Control `tc(sendbreak|drain|flush|flow)`
  - Windows `ioctl`
  - Pseudo `(posix_open|grant|unlock)pt`, `ptsname`

Terminal I/O is important if you want to control its aspects, for example, to provide input without an echo or control the size of the console window. Pseudo-terminals are important for various tools like *ssh*.

References and Resources

References and Resources

Now a few slides about references and resources for further reading.

Resources (Construction)

- Windows System Programming, Fourth Edition (Appendix B)
- Programming with POSIX Threads
- The Linux Programming Interface
- Hands-On System Programming with Linux
- Linux System Programming Techniques
- Advanced Programming in the UNIX Environment
- Effective TCP/IP Programming
- UNIX Systems Programming
- Low Level X Window Programming
- Wayland Architecture / Wayland Book

I also compiled a list of books useful for studying various Linux API classes from a software construction perspective. Of all these books that are listed, most of them were at least partially read, and some were read from cover to cover. Windows system programming book is included because it provides a helpful appendix comparing Windows API with Linux API (recall natural transformation). One other book, The Linux Programming Interface, stands out and provides an initial reference for this training classification part. If you are interested in windowing and graphics, I put two such books, too, including the Wayland protocol and architecture, a replacement for the original X11 one.

Wayland Architecture
https://wayland.freedesktop.org/architecture.html

Wayland Book
https://wayland-book.com/

Resources (Postconstruction)

- WinDbg Help / WinDbg.org (quick links)
- DumpAnalysis.org / SoftwareDiagnostics.Institute / PatternDiagnostics.com
- Debugging.TV / YouTube.com/DebuggingTV / YouTube.com/PatternDiagnostics
- Foundations of Linux Debugging, Disassembling, and Reversing
- Foundations of ARM64 Linux Debugging, Disassembling, and Reversing
- Software Diagnostics Library
- Encyclopedia of Crash Dump Analysis Patterns, Third Edition
- Trace, Log, Text, Narrative, Data
- Memory Dump Analysis Anthology (Diagnomicon)

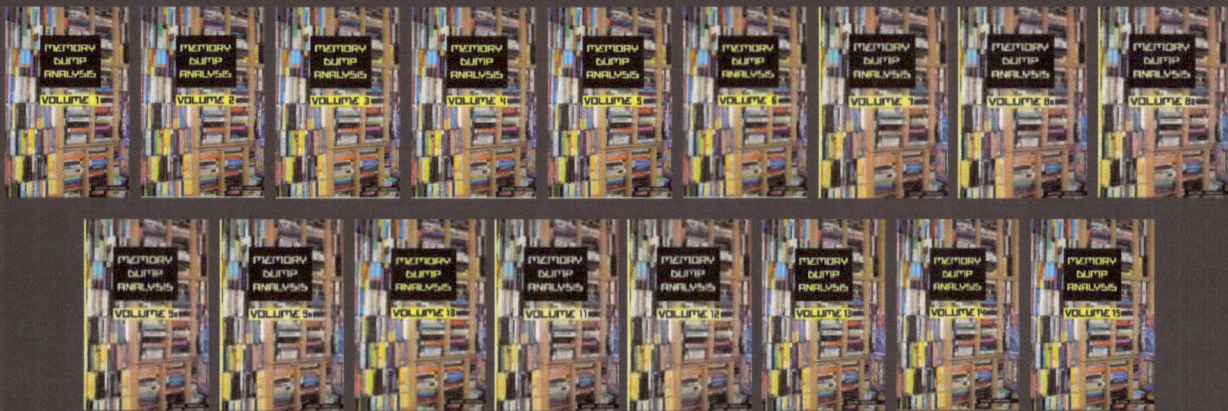

If you don't have experience with assembly language, then the Foundations books teach you assembly language from scratch in the context of GDB. Since I no longer provide them for training attendees, this training includes an assembly language section for both CPU platforms.

WinDbg quick links
http://WinDbg.org

Software Diagnostics Institute
https://www.dumpanalysis.org

Software Diagnostics Services
https://www.patterndiagnostics.com

Software Diagnostics Library
https://www.dumpanalysis.org/blog

Memory Dump Analysis Anthology (Diagnomicon)
https://www.patterndiagnostics.com/mdaa-volumes

Debugging.TV
http://debugging.tv/
https://www.youtube.com/DebuggingTV
https://www.youtube.com/PatternDiagnostics

Foundations of Linux Debugging, Disassembling, and Reversing
https://www.patterndiagnostics.com/practical-foundations-linux-debugging-disassembling-reversing

Foundations of ARM64 Linux Debugging, Disassembling, and Reversing
https://www.patterndiagnostics.com/practical-foundations-arm64-linux-debugging-disassembling-reversing

Encyclopedia of Crash Dump Analysis Patterns, Third Edition
https://www.patterndiagnostics.com/encyclopedia-crash-dump-analysis-patterns

Trace, Log, Text, Narrative, Data
https://www.patterndiagnostics.com/trace-log-analysis-pattern-reference

Resources (Training)

- [Accelerated Linux Core Dump Analysis, Third Edition](#)

- [Accelerated Linux Debugging[4]](#)

- [Accelerated Linux Disassembly, Reconstruction, and Reversing, Second Edition](#)

- [Accelerated Software Trace Analysis, Revised Edition, Part 1: Fundamentals and Basic Patterns](#)

This slide shows various Linux courses I developed over time. We mentioned some ADDR patterns from the *Accelerated Linux Disassembly, Reconstruction, and Reversing* course.

Accelerated Linux Core Dump Analysis, Third Edition
https://www.patterndiagnostics.com/accelerated-linux-core-dump-analysis-book

Accelerated Linux Debugging[4]
https://www.patterndiagnostics.com/accelerated-linux-debugging-4d

Accelerated Linux Disassembly, Reconstruction, and Reversing, Second Edition
https://www.patterndiagnostics.com/accelerated-linux-disassembly-reconstruction-reversing-book

Accelerated Software Trace Analysis, Revised Edition, Part 1: Fundamentals and Basic Patterns
https://www.patterndiagnostics.com/accelerated-software-trace-analysis-part1

Resources (Category Theory)

Applied category theory books that have chapters explaining category theory:

- Conceptual Mathematics: A First Introduction to Categories
- The Joy of Abstraction: An Exploration of Math, Category Theory, and Life
- Category Theory for Programmers
- Categories for Software Engineering
- An Invitation to Applied Category Theory: Seven Sketches in Compositionality
- Life Itself: A Comprehensive Inquiry Into the Nature, Origin, and Fabrication of Life
- Category Theory for the Sciences
- Conceptual Mathematics and Literature: Toward a Deep Reading of Texts and Minds
- Diagrammatic Immanence: Category Theory and Philosophy
- Mathematical Mechanics: From Particle To Muscle
- Memory Evolutive Systems; Hierarchy, Emergence, Cognition
- Mathematical Structures of Natural Intelligence
- Sheaf Theory Through Examples
- Visual Category Theory

As I promised when introducing category theory, this slide contains references to applied category theory books. I also highlighted three books that had some influence on me. For example, they provided a deeper understanding of applied aspects that resulted in a better understanding of mathematical aspects.

Category Theory for Programmers
https://github.com/hmemcpy/milewski-ctfp-pdf

Visual Category Theory
https://dumpanalysis.org/visual-category-theory

www.ingramcontent.com/pod-product-compliance
Lightning Source LLC
Chambersburg PA
CBHW050836220326

41598CB00006B/375